Computational Molecular Science and Engineering Forum 2018

Core Programming Area at the 2018 AIChE Annual Meeting

Pittsburgh, Pennsylvania, USA
28 October - 2 November 2018

ISBN: 978-1-5108-7615-6

Printed from e-media with permission by:

Curran Associates, Inc.
57 Morehouse Lane
Red Hook, NY 12571

Some format issues inherent in the e-media version may also appear in this print version.

Printed by Curran Associates, Inc. (2019)

For permission requests, please contact AIChE
at the address below.

AIChE
120 Wall Street, FL 23
New York, NY 10005-4020

Phone: (800) 242-4363
Fax: (203) 775-5177

www.aiche.org

Additional copies of this publication are available from:

Curran Associates, Inc.
57 Morehouse Lane
Red Hook, NY 12571 USA
Phone: 845-758-0400
Fax: 845-758-2633
Email: curran@proceedings.com
Web: www.proceedings.com

TABLE OF CONTENTS

Some abstracts in this book refer to images that were not made available by the publisher

541272 MoSDeF: Molecular Simulation and Design Framework

Abstract:

Here we present MoSDeF [1], a suite of extensible Python tools designed to facilitate system initialization and atom-typing for molecular dynamics simulations in a way that is programmatic, scriptable, and reproducible. MoSDeF is designed around two key pieces of software, mBuild [2,3] and Foyer [4,5]. mBuild is designed to facilitate the generation of starting configurations through the definition of reusable, interchangeable "compounds" that can be connected together in a programmatic fashion. mBuild supports parameterized structures through generative modeling, allowing complex systems to be hierarchically constructed from smaller subunits. This allows users to declaratively express repetitive structures, e.g. polymer chains, crystal structures, planar or spatial tiling, and parameterize affine transformations applied to subcomponents of a composite component. mBuild also includes a robust library for the generation of crystal structures. Foyer is a general tool for applying force fields to molecular systems (i.e., atom-typing) that does not rely upon rigid rule hierarchies to atom-type a system that is commonly found in most simulation software. Foyer is designed such that it separates the engine that evaluates the atom-typing rules from the rules themselves, making it easier to debug, extend, and disseminate force fields. Foyer provides a force field and simulator agnostic schema for defining parameter usage that relies upon SMARTS [6] based annotations of chemical context, providing a single documentation of parameter usage that is both human and machine readable and stored with the parameters themselves in a single XML document. mBuild and Foyer are able to output fully parameterized structures to various file formats used by common simulation engines (e.g. GROMACS, LAMMPS, HOOMD-Blue etc.). As such, MoSDeF provides a flexible, programmatic, yet clear approach to encapsulate the routines and parameters used when initializing and performing molecular simulations, which increases transparency and reproducibility of simulation results.

[1] https://github.com/mosdef-hub
[2] C. Klein, J. Sallai, T. J. Jones, C. R. Iacovella, C. McCabe, and P. T. Cummings, "A Hierarchical, Component Based Approach to Screening Properties of Soft Matter," in Foundations of Molecular Modeling and Simulation, 2016, pp. 79–92.
[3] https://github.com/mosdef-hub/mbuild
[4] Iacovella, C. R.; Sallai, J.; Klein, C.; Ma, T. In Idea Paper: Development of a Software Framework for Formalizing Forcefield Atom-Typing for Molecular Simulation, 4th Workshop on Sustainable Software for Science: Practice and Experiences (WSSSPE4), 2016.
[5] https://github.com/mosdef-hub/foyer
[6] http://www.daylight.com/dayhtml/doc/theory/theory.smarts.html

541273 Advanced Sampling Using Ssages: Basics, Practical Tips and More

Abstract:

Advanced sampling techniques, in particular ones which can compute free energies and reactive pathways, are an important new development in molecular simulations. Stretching back 20 years to their origins in multicanonical sampling and the first “flat histogram” methods, significant improvements have been made that make the calculation of complex free energy landscapes possible. Indeed, these techniques have been applied to many disparate problems, such as material constant calculations, determination of native protein folding states and pathways, equilibrium phase diagrams, and self-assembling materials.

In this short course, we will explain the theoretical underpinnings behind common sampling techniques, and discuss the relative merits of different techniques for a given problem. This will be facilitated through demonstrations using our powerful sampling suite SSAGES [1] (pronounced “sages”). SSAGES encompasses a variety of advanced sampling techniques for determining free energy landscapes and reactive pathways. Specifically, we will discuss the application of SSAGES to obtain elastic properties, polymer conformational energies, and nanoparticle potentials of mean force. We will walk through the definition of collective variables necessary for advanced sampling, discuss what makes a good collective variable, and how to address common issues encountered in the convergence of advanced sampling methods. We will also perform hands-on examples to give participants a feel for how SSAGES works, and how to understand advanced sampling methods in general.

1. Sidky, Hythem, Yamil J. Colón, Julian Helfferich, Benjamin J. Sikora, Cody Bezik, Weiwei Chu, Federico Giberti et al. "SSAGES: Software Suite for Advanced General Ensemble Simulations." *The Journal of chemical physics* 148, no. 4 (2018): 044104.

541274 Design Problems with Biomolecules

Abstract:

This session will highlight tools, tips, and trouble-shooting biomolecular simulations. We will introduce CharmGUI, Glycam.org, Amber Leap, and H++, which are popular tools to build protein, lipid, and/or carbohydrate-based systems with appropriate protonation states, end groups, etc. We will also discuss force fields parametrized specifically for these systems as well as what to do when your ligand is not included in the force field. We will share how we diagnosed and corrected errors in creating several biomolecular systems. Last, we will practice extracting knowledge from biomolecular simulations, determining driving forces in thermodynamically-driven processes.

541275 GOMC: GPU Optimized Monte Carlo

Abstract:

GPU Optimized Monte Carlo (GOMC) is open-source software (https://github.com/GOMC-WSU) for simulating many-body molecular systems using the Metropolis Monte Carlo algorithm. It supports simulations in a variety of ensembles, which include canonical, isothermal-isobaric, grand canonical, and Gibbs ensemble. This allows GOMC to be used to study vapor-liquid and liquid-liquid equilibria, adsorption in porous materials, surfactant self-assembly, and condensed phase structure for complex molecules. GOMC supports a variety of all-atom, united atom, and coarse grained force fields such as OPLS, TraPPE, Mie, and Martini. The software has been written in object oriented C++, and uses OpenMP and NVIDIA CUDA to allow for execution on multi-core CPU and GPU architectures.

It this workshop, we demonstrate how to use GOMC to predict vapor-liquid equilibria of pure components and mixtures using the Gibbs ensemble Monte Carlo algorithm, and adsorption isotherms for gas molecules in porous materials using grand canonical Monte Carlo. Typical workflows for creating input to GOMC, and analyzing output, will be presented. Execution of GOMC on multi-core CPUs and GPUs will be discussed.

541276 Managing Data Spaces, Performing MD, and Analyzing Trajectories with Signac, HOOMD-Blue, and Freud

Abstract:

We demonstrate how to efficiently develop a complete, integrated, and reproducible simulation workflow from the setup of a parameter space to the execution of simulations and the post-processing of output data for analysis and visualization, with Signac, HOOMD-blue, and Freud. Signac [1] provides a simple interface to map parameter spaces to simulation data, manage and search data and metadata, and develop scalable computational workflows. Signac-flow chains and executes data space operations and can easily submit workflows to high-performance clusters. HOOMD-blue [2] performs molecular dynamics and hard particle Monte Carlo simulations of particles fast on GPUs. Freud [3] analyzes particle simulation trajectories and computes order parameters, identifies clusters, and produces potential of mean force and torque plots. All these tools are Python libraries and work well together in a cohesive workflow, but each may also be used on its own. Signac provides a command line interface to ease integration of non-python tools in the workflow. The HOOMD-blue simulation engine may be used on its own, or with other workflow and analysis tools. Freud provides generic analysis techniques that can be applied to particle trajectory data from other simulation tools or experimental data. In this workshop, we will introduce the basics of each package and walk attendees through a complete integrated simulation workflow that combines their strengths.

[1] https://glotzerlab.engin.umich.edu/signac
[2] https://glotzerlab.engin.umich.edu/hoomd-blue/
[3] https://glotzerlab.engin.umich.edu/freud/

541170 Physically Validating Molecular Simulation Results

Abstract:

We present a number of ways that users can test whether their molecular simulation results and workflows properly obey the laws of physics they are intended to simulate. We include a number of tests of different complexity, ranging from simple post-processing analysis to more involved tests requiring additional simulations. These tests can significantly increase the reliability of MD simulations by catching a number of common simulation errors violating physical assumptions, such as integrators that do not conserve energy, deviations from the correct physical ensemble, and lack of ergodicity between degrees of freedom. To make it as easy as possible to apply these tests, we have developed an open-source and platform-independent Python library (https://physical-validation.readthedocs.io), and we will cover in this workshop how to use these tools.

536546 Understanding Confinement Effects in Mixed Ionic Liquid Systems: Insights from Molecular Dynamics Simulations

Abstract:

Molecular simulation has provided unique insights into the structure and dynamics of ionic liquids confined in microporous electrodes, which hold promise for their application as electrolytes in electrochemical capacitors, or supercapacitors. To date, however, most studied have relied on relatively simple and idealized systems, i.e. neat ionic liquids confined in single pores of regular geometry with no pore size dispersity. Real systems include any number of non-idealities, such as non-uniform pore sizes and other electrolytes or molecular species mixed in the fluid. Recent experimental results report complex behavior of mixtures of ionic liquids, with some beneficial and some detrimental features. Such results imply systems could be designed to exploit some of these features which perhaps mitigating some downsides. To this end, we investigate the possibility that ions may partition themselves in porous systems via their own molecular sizes. Using classical molecular dynamics of a system including the ionic liquids EMIM-Tf2N and HMIM-Tf2N in multiple porous electrodes of different sizes, we demonstrate that ions exhibit this partitioning behavior: smaller ions preferentially fill the smaller pores; likewise, larger ions preferentially fill the larger pores. The impacts of these findings on device performance, including rate handling capabilities, supercapacitor power density, and cycling capabilities, are discussed.

532976 Modulating the Self-Assembly of 1-n-Dodecyl-3-Methylimidazolium Octylsulfate Biamphiphilic Ionic Liquid

Abstract:

One of the reasons ionic liquid solvents have gained major interest in chemical industries as a green alternative is due to their strong ability to solvate both polar and nonpolar species simultaneously along with their non-volatile nature owing to their negligible vapor pressure. Ionic liquid with high amphiphilic character and mesoscale nanosegregated structures are attractive for biocatalytic chemical processes. It has been established that potential candidate for such applications is highly dependent on the morphological heterogeneities of polar-nonpolar domains. It is believed that precise control over such molecular level segregation can be obtained by modulating the competition between the electrostatic interactions of the ionic moieties in ionic liquid and short-range attractive interactions between the nonpolar regions. Previously, we demonstrated that the ionic liquid morphology can be tuned from sponge-like to layer-like with the increase in the cationic alkyl chain with the contributions to nonpolar domain formation from both the ions. In fact, 1-n-dodecyl-3-methylimidazolium [C_{12}mim] octylsulfate [C_8SO_4] portrayed nanoscale structuring (polarâ€"nonpolarâ€"polar) over multiple length scales. In this work, we employ molecular dynamics simulations to investigate change in the morphology of [C_{12}mim] [C_8SO_4] ionic liquid using two approaches: (a) exposing the ionic liquid to vacuum; (b) mixing the ionic liquid with molecular solvents of varying polarity, that includes water (polar) and 1-octanol (amphiphilic). The characterization of structural features will be discussed in terms of structure factors, number densities, order parameters, and instantaneous simulation snapshots.

530535 Probing Surfactant-Nanoparticle Interactions at Fluid-Fluid Interfaces

Abstract:

An emulsion is a type of colloid that consists of one liquid being dispersed in another immiscible liquid. The interface between the two immiscible phases can be stabilized by both surfactants and nanoparticles. Upon sufficient coverage of nanoparticles, the presence of a tightly packed colloidal network structure at the fluid-fluid interface provides enhanced mechanical rigidity and viscosity to the interfacial film. The surfactant-nanoparticle interactions play a critical role in determining the stability of the interfacial film. Our research aims at probing the fundamental mechanisms governing the surfactant-nanoparticle interactions at fluid-fluid interfaces and the consequence of surfactant-nanoparticle interactions on surfactant partitioning at such an interface. We will measure interfacial tension and elasticity to quantify the response of the interface to these interactions. The effect of particle morphology and wettability on the surfactant-nanoparticle interactions will be determined. Hydrophilic, partially hydrophobic, and hydrophobic silica nanoparticles will be used to quantify the effect of wettability on surfactant-nanoparticle interactions. The influence of the mode of nanoparticle addition on surfactant partitioning at the fluid-fluid interface will be determined. Non-ionic and ionic surfactants will be used to determine the influence of surfactant type on interactions with nanoparticles. To complement the experimental observations, molecular simulation results will be presented to shed light on the extent of interfacial adsorption of surfactant and nanoparticles. In addition, it will be demonstrated how the presence of such nanoparticles modify the interfacial structuring of the two liquids.

537432 Force Field Parameters for Hydrogen, Oxygen, and Nitrogen to Study Complex Phase Equilibria and Interfacial Reactions

Abstract:

Hydrogen, oxygen, and nitrogen play a vital role in phase equilibria and catalytic reactions like water splitting, fuel cells, and batteries. However, the dynamics of complex electrolyte and metal-solution-gas interfaces is difficult to monitor by imaging and spectroscopy. Molecular dynamics simulations are capable to provide missing insights and here we present new bonded potentials for O_2, H_2 and N_2 as part of the Interface force field (IFF). A combination of harmonic bond stretching and Lennard-Jones parameters (with 12-6 and 9-6 options) is employed and the parameters are fully compatible with existing parameters for organic and inorganic compounds. The force field reproduces equilibrium bond lengths, vibrational features, liquid densities below the boiling point, vaporization energies, and solvation energies (Henry constants) in water at room temperature using standard water models (SPC, TIP3P) in quantitative (<1%) agreement with experimental measurements. The models thus reproduce chemical bonding, structures, and energies as required for a consistent Hamiltonian and significantly broaden the scope of simulation capabilities for gases and solid/liquid/gas electrolyte interfaces. The parameters can be added into many force fields including IFF, CVFF, PCFF, AMBER, CHARMM, OPLS-AA, COMPASS and GROMACS. The same parameterization protocol can also be applied to any other gas molecules.

532142 Investigating the Effect of Concentration on the Interaction of Electrolytes with Interfaces

Abstract:

The self-assembly of proteins, and nanoparticles is a complex phenomenon driven by multi-scale interactions between various interfaces. The forces between these interfaces can change with varying environmental factors like pH, solute concentration, and electrolyte concentration. All-atom molecular simulations of these systems provide key, molecular-level insights about these forces. However, given the high computational cost of simulating "realistic" systems most simulations are performed in dilute electrolyte environments. Thus, these simulations are unable to capture the effect of the ions even though several interfacial adsorption and assembly processes are driven by ions.[1]

In this talk, I will highlight recent results that show that when the electrolyte ions are not accounted for in the sampling scheme, they can limit the sampling of the adsorption of peptides onto interfaces.[2] Then, I will discuss the application of recently introduced advanced sampling techniques (metadynamics) to alleviate this problem.[3,4] Finally, as a proxy for other interfaces, I will highlight how electrolyte adsorption to a mica surface changes with changing electrolyte concentration. This has implications for both self-assembly and adsorption processes since most macroscopic theories that model these processes only account for dilute concentration limits.[5]

REFERENCES

(1) Ma, X.; Zhang, S.; Jiao, F.; Newcomb, C. J.; Zhang, Y.; Prakash, A.; Liao, Z.; Baer, M. D.; Mundy, C. J.; Pfaendtner, J.; et al. Tuning Crystallization Pathways through Sequence Engineering of Biomimetic Polymers. *Nat. Mater.* **2017**, *16* (7), 767–774.

(2) Prakash, A.; Sprenger, K. G.; Pfaendtner, J. Essential Slow Degrees of Freedom in Protein-Surface Simulations: A Metadynamics Investigation. *Biochem. Biophys. Res. Commun.* **2017**, *498* (2), 274–281.

(3) Pfaendtner, J.; Bonomi, M. Efficient Sampling of High-Dimensional Free-Energy Landscapes with Parallel Bias Metadynamics. *J. Chem. Theory Comput.* **2015**, *11* (11), 5062–5067.

(4) Prakash, A.; Fu, C. D.; Bonomi, M.; Pfaendtner, J. Biasing Smarter, Not Harder, By Partitioning Collective Variables Into Families. *Under Rev.* **2018**.

(5) Prakash, A.; Pfaendtner, J.; Chun, J.; Mundy, C. J. Quantifying the Molecular-Scale Aqueous Response to the Mica Surface. *J. Phys. Chem. C* **2017**, *121* (34), 18496–18504.

536162 Screening Structure-Property Relationships in Lubricating Monolayer Films through Molecular Dynamics Simulation

Abstract:

Nanoscale devices featuring surfaces in sliding contact exhibit high surface area-to-volume ratios and surface energies that promote stiction and surface wear, leading to device failure and design limitations. Alkane-based surface monolayers have been shown to reduce stiction while yielding low coefficients of friction and as such have been proposed as a potential lubrication solution to minimize such limitations. The tribological performance of monolayer lubricants is closely linked to their chemistry, with, for example, terminal group functionality and chain length shown to alter adhesion and frictional performance of monolayer films. However, only a small subset of the vast parameter space available to monolayer chemistry has been examined to-date. Wide screening of various chemistries is thus desirable to obtain a thorough understanding of the links between chemistry, structure, and optimal tribological properties.

To this end, we have developed the molecular simulation and design framework (MoSDeF) software suite [1] that provides scripted, flexible, and reusable routines to facilitate screening over chemical/structural parameter spaces. Using MoSDeF, screening of functionalized monolayer films featuring sixteen distinct terminal groups and five chain lengths has been performed. Both single- and mixed-component systems (where the top and bottom monolayers feature different chemistries) have been studied, with the goal of identifying trends between monolayer chemistry and tribology to aid in the design of better-lubricating films. Structure-property relationships are obtained by assigning a series of chemical descriptors to each terminal group and examining correlations with adhesion and coefficient of friction. We detail how characteristics such as asphericity and hydrophobicity influence the tribology of the monolayer films and provide insight into the characteristics of an "ideal" monolayer chemistry. These relationships are further related to underlying friction mechanisms via analysis of monolayer structural properties such as orientational chain order, average tilt angle, and gauche defects.

References

[1] "MoSDeF" [Online]. Available: https://github.com/mosdef-hub.

536478 Surface Diffusion of Large Molecules: A Computational Study

Abstract:

Surface Diffusion of Large Molecules: A Computational Study

Kutay B. Sezginel and Christopher E. Wilmer

Department of Chemical and Petroleum Engineering, University of Pittsburgh, 3700 Oâ€™Hara Street, Pittsburgh, Pennsylvania 15261, United States

Controlling molecular motion on surfaces is one of the first steps towards bottom-up construction of nanoscale machines. This process is commonly utilized in cells to transport molecular cargo: enzyme molecules are moved along protein filament tracks converting chemical energy into mechanical work. Careful design of the molecular structure and selection of the appropriate surface can enable precise manipulation of the molecular diffusion. Several strategies have been shown to greatly influence diffusion such as molecular functionalization and adjusting the orientation of the molecule with respect to the substrate lattice. Although these strategies are useful, to achieve a nanoscale understanding and control over the dynamics of molecular motion on surfaces it is crucial to develop theoretical models and utilize modern computational tools. Developing a robust computational method is crucial to get further insights and deeper understanding of the molecular surface diffusion which then can be used to design molecules to tailor diffusion properties such as speed and directionality.

In this work, we investigated the diffusion of 9 large organic molecules on a Cu (110) surface. The molecules were selected from available experimental diffusion data in the literature. First, we reviewed different modeling strategies to qualitatively reproduce experimental diffusion results using molecular dynamics (MD) simulations. After developing a robust method, we then calculated the speed and the directionality of the motion of these molecules. Analyzing the MD trajectories, we studied the effect of long jumps (random jumps between nearest neighbor sites) and calculated hopping rate for each molecule. By comparing the nature of diffusion of each molecule we present insights on how to optimize directionality and speed of the motion.

534401 Meso-Scale Modeling of PNIPAM Brushes

Abstract:

Poly(N-isopropylacrylamide) (PNIPAM) is a temperature sensitive polymer with a lower critical solution temperature (LCST) of ~305 K. Below its LCST, PNIPAM is in a coil-state and above the LCST it is in a globule-state. Recently, we have developed a new coarse-grained (CG) model of PNIPAM, which can predict its LCST with a reasonable accuracy. Here, we have utilized this new CG model to perform molecular dynamics (MD) simulations of PNIPAM grafted planar brush structures. Our study is focused on the effect of temperature and grafting density on the coil-to-globule transition in PNIPAM. The CG MD simulations were carried out for ~1 μs at 295 K and 315 K, which is below and above the LCST of PNIPAM, respectively. Simulation trajectories were analyzed for structural and dynamical properties of polymer chains and water.

527499 Molecular Simulation of Ionic Polyimides and Ionic Liquid Composite Membranes for Gas Selectivity and Adsorption

Abstract:

Molecular Simulation of Ionic Polyimides and Ionic Liquid Composite Membranes for Gas Selectivity and Adsorption

Asghar Abedini, Joanna Szala-Bilnik, Ellis Crabtree, Jason E. Bara, and C. Heath Turner*

Abstract:

The reduction and control of greenhouse gas emissions is a priority of many industries due to its detrimental environmental and atmospheric effects. Carbon dioxide is one of the most well-known greenhouse gases with many industrial emission sources. To combat emission, several different carbon capture techniques for pre-combustion and post-combustion have been successfully applied.[1] Most carbon capture processes still depend on amine based solvents, such as monoethanolamine (MEA), but these solvents exhibit multiple disadvantages (e.g., corrosion, volatility, toxicity, and high-energy demand for recovery).[2]

Ionic polyimides (i-IPs) are a new class of polymers for membrane-based CO_2 separations with promising pre-combustion, gas sweetening, and CO_2 post-combustion capture applications (Figure 1).[3,4] This novel class of polymers, like poly(ionic liquids)s (PILs),[5,6] contain ionic liquid species in the backbone of the monomer. However, the combination of an organic linker and ionic liquid (IL) in the i-IP monomer structure provides an additional degree of structural and chemical design in the polymer system. In this study, prototype composite structures are modeled that contain i-IPs with pyromellitic dianhydride (PMDA) as an organic linker. The role of different ILs is investigated, with $[C_4mim^+]$ as the cation in combination with one of the following anions: bis(trifluoromethylsulfonyl)imide ($[Tf_2N^-]$), tetrafluoroborate ($[BF_4^-]$), and hexafluorophosphate ($[PF_6^-]$).

We analyze the detailed structural changes (e.g., fractional free volume, pore size distribution, surface area, etc.) of the ionic polyimides and their composites (i-PI+IL) during CO_2 adsorption using a combination of molecular dynamic (MD) calculations and grand canonical Monte Carlo (GCMC) simulations. The extracted results shows that in $[BF_4^-]$-based materials an addition of 50% IL can increase CO_2/CH_4 selectivity by 16%, and in $[PF_6^-]$-based materials the selectivity is increased by 36%. Overall, it is found that the $[BF_4^-]$-based system shows higher CO_2/CH_4 gas separation (potential pre-combustion applications), while the $[Tf_2N^-]$-based i-PI+IL composites shows higher CO_2/N_2 selectivity performance (potential post-combustion applications). Higher gas solubility correlates with higher theoretical surface area values and a higher fraction of larger pore sizes, while the FFV is not a sensitive indicator of the solubility.

Figure 1. Representative monomer structure of a neat i-IP. The anion shown is $[BF_4^-]$. Specific nitrogen sites of the i-IP are labeled for reference, as well as the head (C_H) and tail (C_T) designation.

References:

(1) Martín, C. F.; Stöckel, E.; Clowes, R.; Adams, D. J.; Cooper, A. I.; Pis, J. J.; Rubiera, F.; Pevida, C. Hypercrosslinked organic polymer networks as potential adsorbents for pre-combustion CO_2 capture. *Journal of Materials Chemistry* **2011**, *21*, 5475.

(2) Abdul Halim, H. N.; M. Shariff, A.; Tan, L. S.; Bustam, M. A. Mass Transfer Performance of CO_2 Absorption from Natural Gas using Monoethanolamine (MEA) in High Pressure Operations. *Industrial & Engineering Chemistry Research* **2015**, *54*, 1675-1680.

(3) Abedini, A.; Crabtree, E.; Bara, J. E.; Turner, C. H. Molecular Simulation of Ionic Polyimides and Composites with Ionic Liquids as Gas-Separation Membranes. *Langmuir* **2017**, *33*, 11377-11389.

(4) Mittenthal, M. S.; Flowers, B. S.; Bara, J. E.; Whitley, J. W.; Spear, S. K.; Roveda, J. D.; Wallace, D. A.; Shannon, M. S.; Holler, R.; Martens, R.; Daly, D. T. Ionic Polyimides: Hybrid Polymer Architectures and Composites with Ionic Liquids for Advanced Gas Separation Membranes. *Industrial & Engineering Chemistry Research* **2017**, *56*, 5055-5069.

(5) Morozova, S. M.; Shaplov, A. S.; Lozinskaya, E. I.; Mecerreyes, D.; Sardon, H.; Zulfiqar, S.; Suárez-García, F.; Vygodskii, Y. S. Ionic Polyurethanes as a New Family of Poly(ionic liquid)s for Efficient CO_2 Capture. *Macromolecules* **2017**, *50*, 2814-2824.

(6) Qian, W.; Texter, J.; Yan, F. Frontiers in poly(ionic liquid)s: syntheses and applications. *Chem Soc Rev* **2017**, *46*, 1124-1159.

522245 Pablo Debenedetti: Recollections from the Past 36 Years

Abstract:

In this talk, I would like to provide some personal reminiscences related to scientific, professional, and personal interactions I had with Pablo Debenedetti since we first met as graduate students at MIT. I will highlight how his work has evolved over this time and the deep impact he had on a number of different fields in which I had the opportunity to work with him: supercritical fluids, metastability, liquids and glasses, advanced sampling algorithms, proteins in solution, molten metals, the liquid-liquid phase transition in supercooled water, and nucleation of salts from aqueous solution.

537180 Pressure: The Neglected Variable in High Pressure Processing

Abstract:

High pressures are required for many processes in the chemical, oil and gas, pharmaceutical and other industries. Examples can be found in many heterogeneous gas reactions, in the production of pharmaceuticals, and in the synthesis of high pressure solid phases of materials with desired properties, such as semiconducting or superconducting behaviour. In some cases pressures of thousands or tens of thousands of bar are necessary. For some applications even higher pressures may be needed, approaching one million bar or more; such pressures can be achieved only in diamond anvil cells for very small samples, and are not practical at present on an industrial scale. Even when possible, achieving such high pressures is very energy-intensive, expensive, and harmful to the environment. In this talk an alternative approach to the production of high pressures will be considered, based on the very strong attractive force fields exerted by some solid substrates. Such an approach is not energy-intensive, environmentally damaging or expensive, and so could lead to a promising new technology. It is mainly limited by our current lack of fundamental understanding of the atomic level processes involved.

There is an abundance of anecdotal evidence that nano-phases adsorbed on solid substrates or within nano-porous materials can exhibit high pressures as a result of the confinement[1,2]. For example, phase changes and chemical reactions that only occur at high pressures in the bulk phase occur in the adsorbed film or confined phase at bulk phase pressures that are orders of magnitude lower. The pressure in the film is different in different directions; for simple surface geometries there is a pressure normal to the surface of the substrate, and one parallel to the walls components (tangential pressure).

For simple fluids in pores that are up to a few nanometers in width, molecular simulations show that both the normal and tangential pressures can be locally very high (thousands or tens of thousands of bars, for example) in the film, even though the bulk phase in equilibrium with the pore is at a pressure of one bar or less. The cause of these high in-film pressures will be discussed, and where possible comparison with experimental results will be made[3].

When the molecules in the confined nano-phase react with each other chemically, or with the pore walls, it may be possible to achieve even higher tangential pressures, in the megabar range. Evidence for this is provided by recent experiments on sulfur (an insulator at ambient conditions) in narrow single-walled carbon nanotubes, carried out by Kaneko and coworkers[4]. They find that the sulfur atoms within the pore covalently bond to form a one-dimensional phase that is metallic. In the bulk phase sulfur forms a metallic phase only at pressures above 95 GPa. In our recent molecular dynamics simulations of this system[5] we find that the sulfur atoms are covalently bonded in the pore and that they experience tangential pressures in excess of 100 GPa as a result of the strong confinement. In a second example, the nitric oxide dimer reaction, 2NO = $(NO)_2$, the dimer molecules interact strongly ('chemically') with the pore wall, leading to a 100% yield of the dimer in the pores, in contrast to a yield of less than 1% in a gas phase at the same thermodynamic conditions. Again the tangential pressures in the reacting phase near the pore walls is found to be in the megabar range.[6]

[1] Yun Long, Jeremy C. Palmer, Benoit Coasne, Małgorzata Śliwinska-Bartkowiak and Keith E. Gubbins, "Pressure enhancement in carbon nanopores: A major confinement effect", *Physical Chemistry Chemical Physics*, **13**, 17163-17170 (2011).

[2] Yun Long, Jeremy C. Palmer, Benoit Coasne, Małgorzata Śliwinska-Bartkowiak, George Jackson, Erich A. Müller and Keith E. Gubbins, "On the Molecular Origin of High Pressure Effects in Nanoconfinement: Effects of Surface Chemistry and Roughness", *Journal of Chemical Physics*, **139**, 144701 (2013)

[3] M. Śliwinska-Bartkowiak, H. Drozdowski, M. Kempinski, M. Jazdzewska, Y. Long, J.C. Palmer and K.E. Gubbins, „Structural Analysis of the Behavior of Water Adsorbed in Activated Carbon Fibers", *Physical Chemistry Chemical Physics*, **14**, 7145-7153 (2012).

[4] Y. Fujimori, A. Morelos-Gómez, Z. Zhu, et al., "Conducting Linear Chains of Sulphur Inside Carbon Nanotubes", *Nature Comm.*, 4, 3162 (2013).

[5] Cody K. Addington, J. Matthew Mansell and Keith E. Gubbins, "Computer Simulation of Conductive Linear Sulfur Chains Confined in Carbon Nanotubes", *Molecular Simulation*, **43**, 519-525 (2016).

[6] Deepti Srivastava, Erik E. Santiso and Keith E. Gubbins, "Dimerization of Nitric Oxide: Effect of Confinement in Carbon Nanopores", to be published (2017).

519667 Polymer-Grafted Nanoparticle Membranes with Controllable Free-Volume

Abstract:

Polymer based membranes play a key role in several industrially important gas separation technologies, e.g., removing $CO_2$from natural gas, with enormous economic and environmental impact. Baker advocates the development of novel membrane architecturessince current, pure polymer membranes only offer limited systematic pathways for improvement. Here,we develop a novel hybrid membrane construct comprised entirely of nanoparticles grafted with polymers. These membranes are shown to have broadly tunable separation performance through variations in graft density and chain length. Computer simulations show that the optimal NP packing forces the grafted polymer layer to distort, yielding regions of measurably lower polymer density. Multiple experimental probes confirm that these materials have the predicted increase in â€œpolymer free volumeâ€, which explains their improved separation performance. These polymer-grafted NP materials thus represent a new template for rationally designing membranes with desirable separation abilities, coupled with improved aging characteristics in the glassy state and enhanced mechanical behavior.

524526 On the Thermodynamics of Systems Under the Influence of Gravity

Abstract:

Given Pablo Debenedetti's fondness of, in addition to his nearly unparalleled understanding of, classical thermodynamics, it is therefore fitting to present in his honor on a topic in thermodynamics to which he himself has made important contributions. While there are many examples to choose from, I will limit my discussions here to two aspects of the thermodynamics of systems subject to a gravitational field.

First, I revisit Pablo's earlier work on the thermodynamic stability of fluids under the influence of gravity (1988, *J. Chem. Phys.* **89**, 6881). In that work, the standard (i.e., gravity-free) thermodynamic stability criteria were shown to be unaffected by the presence of gravity. Thus, the single-phase fluid is thermodynamically stable when exposed to a gravitational field only if the standard stability criteria are satisfied everywhere throughout the fluid. While the validity of this conclusion is not in question, I nonetheless present a new derivation of some of the results contained in Pablo's earlier work. This new approach makes use of a novel method for obtaining the extrema of a multivariable function in which the initially independent variables are subject to a given set of constraints, and which also avoids the use of Lagrange multipliers. Furthermore, and because Lagrange multipliers are not required, explicit expressions for the second "constrained derivatives" are obtained. Hence, the nature of the constrained extrema (e.g., maxima or minima) can be readily determined. By extending this constrained analysis from the discrete variable case to the continuum variable limit, thereby yielding constrained functional derivatives, thermodynamic extrema principles (based on the minimization of various thermodynamic potentials subject to appropriate constraints) are then invoked to obtain the conditions of equilibrium and stability for fluids in the presence of a gravitational field.

Second, I reconsider a recent analysis of a solid sphere located within a gravitational field (De Palma and Sormani, 2015, *Am. J. Phys.* **83**, 723). First, after initially resting on a platform, the temperature of the sphere is increased, which serves to raise its center of mass due to the resulting thermal expansion. Next, the top of the sphere is then attached to a taut string while (temporarily) removing the platform. Thus, a subsequent decrease in the temperature of the sphere serves to raise yet again its center of mass. Finally, work can be generated by letting the sphere fall back down to the platform (now returned to its initial location). Consequently, a cyclic process has been created that (apparently) violates the second law of thermodynamics. The claimed violation only appears, according to the authors' analysis, when the internal energy is assumed to be independent of the gravitational field, as is normally done for such systems. No thermodynamic inconsistencies presumably arise if the internal energy is instead chosen to be a function of the strength of the gravitational field. Such a conclusion does not, however, immediately follow from this previous analysis, as certain work interactions during the expansion/contraction of the sphere were not rigorously determined. To show this more clearly, I consider in lieu of the sphere a similar cyclic process generated using an ideal gas in a container with a movable piston (and with a similar net upward movement of the center of mass of the gas). By accounting properly for the pressure variations that develop at equilibrium in a gravitational field, as well as comparing the process to the appropriate Carnot cycle, I show that this cyclic process does not in fact violate any laws of thermodynamics. Thus, the standard assumption of the independence of the internal energy from the gravitational field can still be safely invoked.

539767 Thermodynamics of Ionic Liquid Mixtures

Abstract:

Ionic liquids show promise for a variety of applications due to the ability to tune both their chemical and physical properties. Further design parameters are accessible by mixing ionic liquids with conventional solvents, as well as using mixtures of ionic liquids. Here we will explore both expected and unexpected behavior of mixtures containing ionic liquids, especially focusing on thermophysical properties such as viscosity.

527445 Forward Flux Sampling Using Jumpy Order Parameters

Abstract:

Forward-flux sampling (FFS) is a path sampling technique that has gained increased popularity in recent years, and has been used to compute rates of rare event phenomena such as crystallization, condensation, hydrophobic evaporation, DNA hybridization and protein folding. The popularity of FFS is not only due to its ease of implementation, but also because it is not very sensitive to the particular choice of an order parameter. The order parameter utilized in conventional FFS, however, still needs to satisfy a stringent smoothness criterion in order to assure sequential crossing of FFS milestones. This condition is usually violated for order parameters utilized for describing aggregation phenomena such as crystallization. In this talk, I will present a generalized FFS algorithm for which this smoothness criterion is no longer necessary, and apply it to compute homogeneous crystal nucleation rates in several systems. Our numerical tests reveal that conventional FFS can sometimes underestimate the nucleation rate by several orders of magnitude.

520885 Machine Learning for Design and Detection of Assembly

Abstract:

Nanometer-scale, colloidally-stable particles suspended in a fluid can be driven to assemble into a wide variety of different structures depending on the control parameters of the system and the nature of the effective interparticle interactions. In many cases, the relevant interactions are tunable via external fields, physical or chemical modification of the particle surfaces, or changes in the composition of the suspending solvent. In this talk, we discuss challenges associated with the inverse design of interactions for assembly into a targeted structure, the detection of such a transition, and the opportunities that new machine learning based simulation approaches provide for addressing both.

540773 2018 Outlook for Energy: A View to 2040

Abstract:

The Outlook for Energy is ExxonMobil's global view of energy demand and supply through 2040. It also highlights the dual challenge of providing affordable energy to support prosperity while reducing environmental impacts – including the risks of climate change. We use the Outlook to help inform our long-term business strategies and investment plans.
The Outlook anticipates significant changes through 2040 across the world to boost living standards, reshape the use of energy, broaden access to abundant energy supplies, and accelerate decarbonization of the world's energy system to address the risks of climate change.

- Global energy needs will rise about 25 percent over the period to 2040
- Electricity demand will nearly double in non-OECD countries
- Worldwide electricity from solar and wind will increase about 400 percent
- Natural gas will expand its role, led by growth for electricity generation
- Growth in oil demand will be driven by commercial transportation and chemical needs
- Efficiency gains and growing use of less-carbon intensive energy sources will contribute to a nearly 45 percent decline in the carbon intensity of global GDP

Energy is fundamental to modern life, and as the world's population approaches 9 billion people in 2040, we are challenged to improve living standards everywhere. Understanding the factors that drive the world's energy needs – and likely choices to meet those needs – is the mission of the Outlook. By sharing the Outlook with the public, we hope to broaden that understanding. Energy matters to everyone, and we all play a role in shaping its future.

541543 Energy Decarbonisation Scenarios

Abstract:

With the energy sector being associated with 2/3 of the annual greenhouse gas emissions, decarbonizing the energy is a key step to reduce their level. The National Determined Contributions (NDC's) announced so far by the participating countries do not meet the objectives of the Paris-climate agreement, as far as reaching a peak in emissions in the near future or decreasing to a level that is compatible with the stated long-term targets.

Using different scenarios for energy demand and supply, options for energy sector decarbonisation are evaluated. Meeting the 2-degree target requires a major transformation in the power sector, efficiency improvements – specially in the buildings and in the industry, and a revolution in transport towards fuel and modal shifts. With a significant increase in renewable energy, system integration and optimization will also be key in reducing the cost of the energy transition. Meeting more ambitious climate goals ("well below 2 degrees") will push further the technological innovation and implementation requirements throughout the entire energy value chain. A scenario aiming at zero net-emissions by 2060 is presented, along with a comparison to the 2-degree case.

542625 Fundamental Research Needs to Advance Energy Technologies

Abstract:

The U. S. Department of Energy (DOE), Office of Science, Office of Basic Energy Sciences (BES) supports fundamental research in chemical and materials sciences to provide the foundations for new energy technologies and to support DOE missions in energy, environment, and national security. This presentation will discuss opportunities for fundamental research to impact DOE's energy mission "to catalyze the timely, material, and efficient transformation of the nation's energy system and secure U.S. leadership in energy technologies" with a focus on the way we generate, store and use energy nationally. I will provide an overview of BES strategic planning over the past decade that identified priority research directions for advancing energy applications, highlight key scientific advances in these areas, and discuss some future opportunities for modern science, particularly chemical sciences, to accelerate the transformation of the U. S. energy portfolio.

521547 Enhanced Coil Dimensions of Polyolefins: Effect of Tacticity and Side Chain Structure

Abstract:

The development of new polymeric materials with desirable properties hinges on a solid understanding of structure-property relationships. In this work, we evaluate the relationship between the polymer structure and their coil dimensions and explore new polyolefins based on these relations. Coil dimensions are important features of polymers which affects their performance in various applications, including drug delivery, waste-water treatment, construction material, and engine oils. Coil dimensions are dependent on the composition, topology or connectivity, molecular weights, distributions and modalities. Other contributions include the number, size, and stereo orientation of side chains along the polymer backbone. Thus, controlling these attributes allows us to tailor the coil dimensions of polymers. Understanding the fundamental physics of these systems is vital for their development.

We present a computational approach to evaluate the coil dimensions of polymers in dilute solutions. In the proposed scheme, we calculate the radius of gyration (Rg) of polymer chains using molecular dynamics simulations. We use simulated annealing to ensure the capture of different coil configurations. We demonstrate the validity of the proposed model by comparing its predictions with experimentally known Rg values of stereoregular Î±-olefin polymers as representative systems. We demonstrate that this approach affords the ability to quantify the effect tacticity has on the coil dimensions of polyolefins. We observe that when the polymer chains transition to a bottlebrush structure the effect of tacticity is suppressed. Furthermore, we demonstrate that the side chain steric hindrance plays an important role in the rigidity of the polymer backbone. In addition to tacticity, we use the model to evaluate the effects of polymer composition on the coil dimensions. Combining our model with virtual high-throughput screening techniques, we evaluate the coiling behavior of hundreds of new polymers. Using the screening results, we obtain correlations between the structure of the side chain and the coil dimensions of polymers.

537090 Mechanisms of Halogenated Silane Decomposition on an N-Rich Surface during Atomic Layer Deposition of Silicon Nitride

Abstract:

Atomic layer deposition (ALD) has recently received increasing attention for the growth of high-conformity silicon nitride (SiN) thin films for use in microelectronics, particularly as charge storage layers in vertical-NAND. Plasma enhanced ALD (PEALD) allows SiN deposition at substantially lower temperatures (< 400 °C) with better film properties, compared to thermal ALD. These advantages make PEALD more attractive for ultra large scale integrated circuit (ULSI) device fabrication where the growth of aspect ratio independent and high-quality conformal thin dielectric films is tremendously important. The challenge of PEALD is that Si deposition must occur via a thermal mechanism due to the low temperatures. The PEALD of SiN films involves a two-step cycle: (1) adsorption and decomposition of silicon-containing precursors and (2) nitridation of the Si-rich surface by active N species emanating from the plasma. Halogenated silanes such as hexachlorodisilane, bis(tertiary-butyl-amino)- silane, and dicholorosilane (DCS,
SiH_2Cl_2) have been utilized as Si precursors. Despite efforts directed towards process development, the underlying reaction mechanisms of Si deposition remain uncertain.

Methods to improve growth rate and uniformity have been demonstrated experimentally, but without knowledge of the reaction mechanisms, direct contributions of specific process conditions cannot be explained. Using first-principles density functional theory DFT) calculations, we have examined and identified a novel mechanism for the adsorption and decomposition of DCS on a N-rich SiN surface. Our study predicts that the DCS adsorption and dissociation can occur by overcoming a moderate barrier (â‰ˆ 0.3 eV), lower than the prohibitively large barriers predicted for previously proposed mechanisms. Through a detailed electronic structure analysis of the reaction intermediates, we have also elucidated the principles underlying the reaction mechanism, notably the hypercoordination of Si which permits the facile reaction of molecularly adsorbed DCS with primary and secondary amines on the surface, followed by dissociation releasing protons and Cl anions with subsequent HCl formation. We have examined the same mechanism utilizing alternative precursors and the predicted trends are found to be corroborated with the important properties of the system. Understanding these principles allows us to develop guidelines for processing conditions, such as the importance of maintaining the proper surface composition to facilitate Si precursor adsorption and dissociation. Our study provides insight into the SiN ALD process via chlorosilanes and guidelines to control the deposition for high-quality SiN films and provides a framework for future theoretical studies of surface reactions during ALD.

527747 Effect of Lanthanum Doping on Structure, Electronic and Elastic Properties of Perovskite, Pyrochlore Oxide and Lanthanide Titanates: A First Principles Study

Abstract:

Effect of Lanthanum doping on structure, electronic and elastic properties of perovskite, pyrochlore oxide and lanthanide titanates: A First Principles study

Amar Deep Pathak, Foram Thakkar, Suchismita Sanyal, Hans Geerlings, Arian Nijmeijer

Fuel cells have attracted enormous interest due to their high efficiency to convert chemical energy into electrical energy with very low emission. It can be used in stationary, portable, and transport power generation. It also reduces CO_2 emissions when compared with the conventional hydrocarbon fuels, therefore considered as environmentally clean energy source. Fuel cells are classified based on the electrolyte/electrode material and operating temperature range. Solid oxide electrolyte based fuel cells operate at high temperature and are the most efficient fuel cells. They can handle variety of fuels such as H_2, CO and various hydrocarbons CH_4. [1-2]

Perovskite is a family of compound having ABX_3 type crystal structure, where A and B are cations whereas X is an anion. They exhibit variety of properties due to ferroelectric, dielectric, pyroelectric, and piezoelectric behaviours. There is paramount interest to replace conventional Ni-YSZ cermet electrode with doped perovskite in solid oxide fuel cell [3]. $SrTiO_3$, a member of perovskite family, is a well-known band insulator. Doping of La significantly enhance the electronic conductivity of $SrTiO_3$ [4]. Experimentally, perovskite ($LaTiO_3$), pyrochlore oxide ($La_2Ti_2O_7$) and lanthanide (La_2TiO_5) phase can coexist at higher temperatures (~1500 °C) [5]. Thus, it is essential to investigate the effect of doping in these phases of La and Sr.

In the present study, we have used GGA-DFT to investigate the various La doped structures of $SrTiO_3$ and Sr doped structures of $La_2Ti_2O_7$ and La_2TiO_5. The concentration of dopant is varied between 0-100 % considering all possible configurational structures and its dependence on electronic energy, band gap, fermi level, formation enthalpy and elastic properties are established. A change in conducting behaviour is observed as a function of doping, where a transition from band insulator to a metallic behaviour is observed by doping La in $SrTiO_3$, which is reversed with 100% La replacement in $SrTiO_3$. Such behaviour is not observed in the pyrochlore oxide phase. At the end, we shall report the preferential doping among these phase in a mixture where they all can coexist.

Reference:

[1] Kurokawa, Hideto, et al. "Y-doped $SrTiO_3$ based sulfur tolerant anode for solid oxide fuel cells." *Journal of Power Sources* 164.2 (2007): 510-518.

[2] Baniecki, J. D., et al. "Density functional theory and experimental study of the electronic structure and transport properties of La, V, Nb, and Ta doped $SrTiO_3$." *Journal of Applied Physics* 113.1 (2013): 013701.

[3] Nielsen, Jimmi, et al. "Infiltrated $La_{0.4}Sr_{0.4}$ $Fe_{0.03}Ni_{0.03}Ti_{0.94}O_3$ based anodes for all ceramic and metal supported solid oxide fuel cells." *Journal of Power Sources* 372 (2017): 99-106.

[4] Shanthi, N., and D. D. Sarma. "Electronic structure of electron doped $SrTiO_3$: $SrTiO_{3-\delta}$ and $Sr_{1-x}La_xTiO_3$." *Physical Review B* 57.4 (1998): 2153.

[5] Škapin, Srečo D., Drago Kolar, and Danilo Suvorov. "Phase stability and equilibria in the La_2O_3–TiO_2 system." *Journal of the European Ceramic Society* 20.8 (2000): 1179-1185.

525935 Modelling Solubility of Metal Complexes in Non-Aqueous Media from First Principle Calculations: Application to Redox Flow Cell

Abstract:

Solubility is a very fundamental equilibrium thermodynamical property and needs to be addressed in non-aqueous Redox Flow Cell because the energy density of a cell is directly related with the solubility of redox active species in non-aqueous medium. At present, there is no such efficient way to predict the solubility of transition metal complexes in non-aqueous media from first principle calculations. During solubility, the transition metal complex and the non-aqueous solvent are in equilibrium with each other. At equilibrium condition, one can predict solubility by modelling the activity coefficient at infinite dilution of the redox active species in the non-aqueous solvent. We have used the first principle COSMO calculation followed by post statistical thermodynamical treatment of self-consistent properties of the solute and solvent molecules for our calculations. The transition metal complex and the non-aqueous medium have been treated as a simple binary mixture using â€œLANLâ€ activity coefficient model. We used the new activity coefficient model to explain the duel-solute effect already found in the Vanadium redox flow cell containing Vanadium (III) acetylacetonate ($V(acac)_3$) as redox active species, ionic liquids as supporting electrolytes and Acetonitrile as organic solvent. The present model has been applied to a set of non-aqueous solvents and probable mixtures of them. The solvation mechanism of the metal complexes in those non-aqueous media was obtained from our calculations. We have compared our results with the other experimental and theoretical results wherever they are available to check the accuracy of the new activity coefficient model.

536399 Thermophysical, Interfacial and Transport Properties of Low Global Warming Potential Refrigerants from Molecular Theory and Simulations

Abstract:

The KigaliÂ´s amendment to the Montrealâ€™s Protocol limits the use of hydrofluorocarbons (HFCs) as refrigerants, starting by 2019. This has promoted an active area of research towards the development of low Global Warming Potential (GWP) new refrigerants as well as their implementation in the corresponding applications. Hydrofluoroolefins (HFOs) have been proposed as an environmentally friendlier alternative to third generation HFC refrigerants, but further work on fully characterizing them, and their blends with other compounds, is still required to fully assess their performance to replace the ones in current use. In this work we have used the Statistical Associating Fluid Theory (SAFT) coupled with the Density Gradient Theory (DGT) and Rosenfeld's excess entropy scaling approach to predict the phase behavior, thermophysical, interfacial [1,2] and transport properties of refrigerant blends [3]. Molecular dynamics simulations have also been carried out for a further understanding of the molecular phenomena leading to the macroscopic behavior at selected conditions. We will present and discuss new results concerning third and fourth generation blends containing difluoromethane (R-32) for low to medium temperature refrigeration as well as other alternative attractive blends. Results show that binary mixtures of R-32 with propane (R-290), n-butane (R-600), isobutane (R-600a), trifluoroethane (R-143a) and pentafluoroethane (R-125) exhibit azeotropic phase behaviors at the bulk. The latter prevents any temperature glide across the refrigeration cycle, making it more efficient. In addition, mixtures with n-alkanes also exhibit an aneotropic behavior at the interface where the surface tension demonstrates a strong non-linear dependence on concentration. Molecular dynamic simulations allowed gaining molecular insight of the effect of the interactions at the aneotrope and validating density profiles predicted by the molecular theory. Regarding the transport properties calculations, parameters needed for the excess entropy scaling approach were fitted to saturated viscosity and thermal conductivity data for each of the pure refrigerants and used to predict properties up to high temperatures and pressures in excellent agreement with experimental data. The parameters were then utilized to fully predict transport properties of mixtures under consideration without the need of any extra binary interaction parameter. Mixtures predicted in this work include R-452B and R-454B as low GWP alternative HFO blends for R-32. The obtained results reinforce the validity of using molecular modeling and computational tools for addressing industrial needs, designing and optimizing the corresponding products and processes.

[1] WA Fouad, LF Vega, PCCP 19 (13), 8977-8988 (2017)

[2] WA Fouad, LF Vega, AIChE Journal 64 (1), 250-262 (2018)

[3] WA Fouad, LF Vega, J. Supercritical Fluids 131, 106-116 (2018)

537251 Automatic Construction of Collective Variables for Metadynamics Simulations of Drug Permeation through Lipid Membranes

Abstract:

Understanding the mechanism of permeation of small molecules through cellular membranes and estimating their permeability coefficients are of tremendous importance to the pharmaceutical industry in order to improve design of a novel drug molecule or search for the best drug molecule from an existing library of numerous possible candidates. All atom molecular dynamics simulation is a promising method that can serve this purpose, as atomistic scale features can be effectively resolved. However, the simulation time required in crossing a large energy barrier, typically involved in the permeation process makes this method computationally infeasible to study permeation of drug molecules. This problem can be overcome by using enhanced free energy sampling methods that efficiently sample the entire free energy landscape. A bottleneck in the application of enhanced free energy sampling methods is determining collective variables (CVs) before starting the simulations. This step involves careful assessment of the system to select important degrees of freedom, which requires manual work for investigating every new drug molecule. In this work, we aim to overcome this problem by using time-structure based Independent Component Analysis (tICA). The tICA is an unsupervised machine learning technique to find degrees of freedom that decorrelate slowly in the system. We investigate the permeation of drug molecules through lipid membranes by using transition-tempered metadynamics (TTMetaD) simulations, and automatically construct the CVs for TTMetaD simulations based on the characteristics of the system showing slow dynamics by using tICA. Our findings demonstrate that tICA TTMetaD accelerates efficient exploration of the CV space and reduces the computational time to study drug permeation.

537701 Prediction of the Reactivity of Chemical Compounds with Ozone in Water

Abstract:

The wastewater discharged from chemical plants and pharmaceutical factories contains various chemical compounds. To prevent negative impacts on the environment and human health, these compounds need to be decomposed. Microorganism wastewater treatment is the most common method because of its low cost. However, some chemical compounds, called non-biodegradable compounds, are difficult to decompose by microorganisms. Ozonation is an effective method to decompose non-biodegradable compounds, because of ozoneâ€™s high oxidizability. To apply ozonation for wastewater, reactivity or reaction rate constants of ozone and chemical compounds in the wastewater are critical information. However, the reactivity with ozone differs from compound to compound.

Thanks to the development of density functional theory (DFT) calculation method and the improvement of computer performance, DFT calculation has been more widely applied to reveal chemical reactions or various properties of chemical compounds. Reaction rate constants of chemical compounds are calculated with transition states and activation energies which obtained from DFT calculations. However, optimization to transition state in DFT calculation requires chemical intuition and try-and-error.

In this research, we developed a model to predict reaction rate constants of ozone and chemical compounds in water with DFT calculations of relaxed states. Experimentally measured rate constants for about 100 compounds in previous research by other groups were used to develop the model. These compounds contain various kinds of compounds, such as aromatic compounds, carboxylic acids and alcohols. We applied DFT calculations to these compounds to obtain physical properties, such as highest occupied molecular orbital (HOMO) energies and dipole moments. We also calculated degrees of unsaturation and molecular weights from chemical formulas. With these physical and chemical properties, we constructed a regression model to predict the rate constants. The model can be used to predict the rate constants even by people with little knowledge and experience in quantum chemistry calculation, and it would be useful for judging the applicability of ozone wastewater treatment.

Furthermore, we developed another model to predict chemical oxygen demand (COD) decrease of non-biodegradable compound aqueous solution by ozone treatment. COD is one of the most important indexes of waste water. Our model to predict rate constant was incorporated in to the model to predict COD decrease.

537747 Understanding Amyloid Co-Assembly Bycharge-Complementary Peptides

Abstract:

Peptide self-assembly has long been recognized as a key technology for preparing nanomaterials with desired structures and functions for medical and biotechnology applications. Recently, co-assembly of peptides with complementary partial charges has emerged as a new platform for preparing peptide-based nanomaterials. Two different peptides co-assemble into fibrils rich in Î²-sheet when mixed together but remain random coils when by themselves. It has been hypothesized that peptides with opposite partial charges adopt an alternating arrangement in the fibril, but there is no direct evidence at the molecular level to support this hypothesis. We investigate the co-assembly of three charge-complementary peptide systems: CATCH46 (QQKFKFKFKQQ and EQEFEFEFQE), CATCH44 (KQKFKFKFKQK and EQEFEFEFQE), and CATCH66 (KQKFKFKFKQK and EQEFEFEFQE) using molecular modeling, solid state NMR and standard biophysical measurements. Both simulation and solid-state NMR support the alternating peptide hypothesis but show exceptions in which peptides with the same charge are adjacent to each other. Molecular modeling further shows that the peptides can co-assemble into a Î²-barrel structure; the role of this oligomer in the co-assembly processes is discussed.

532488 Structure and Dynamics of Active Liquid Crystalline Biopolymers

Abstract:

Active materials offer intriguing opportunities for applications ranging from robotics to catalytic systems. Their structure and dynamics, however, are not understood. Our recent efforts have focused on the study of active, liquid crystalline polymers, which exhibit nematic phases that are reminiscent of those found in traditional, small-molecule thermotropic systems. In this lecture, we will provide an overview of recent findings from our group, which suggest that an analogy can be made between suspensions of actin filaments with myosin and dense liquids. In this analogy, the defects that arise in such systems are treated as hard, anisotropic particles. By establishing that connection, it becomes possible to formulate a theory of active systems that could in principle enable design of active materials for engineering applications.

524791 Supercitical Fluids and Spray-Drying: Particles for Phamaceuticals

Abstract:

From 1985-2000, researchers in the Debenedetti group were among the first to demonstrate the use supercritical fluids (SCF) as a solvent or anti-solvent to generate micro-particles of drug molecules, polymers, and proteins with potential applications for drug delivery. These processes involved pressurization of a solution containing the desired molecule, and release of the solution through a nozzle into a miscible phase to form the micro-particles. Some of the findings and approach in this work can be seen in recent developments in spray-drying, a similar unit operation now commonly used in pharmaceutical industry. Advances in the applications of spray drying has enabled extremely low solubility molecules to be viable drug products. Recent work using spray-drying to generate 1) dispersions of amorphous active pharmaceutical ingredients within a polymer phase and 2) protein powders will be discussed.

522581 How to Hit HIV Where It Hurts

Abstract:

Vaccination has saved more lives than any other medical procedure. But, some pathogens have evolved which have defied successful vaccination using the empirical paradigms pioneered by Pasteur and Jenner. I will describe how bringing together theoretical/computational approaches from statistical physics, with immunology is beginning to confront this challenge by developing some of the principles necessary for rational design of vaccines that may eliminate these scourges. One characteristic of many pathogens for which successful vaccines do not exist is that they present themselves in various guises. HIV is an extreme example because of its high mutability, and it continues to wreak havoc, especially in developing countries. I will describe how we developed models to translate data on HIV protein sequences to knowledge of the HIV fitness landscape, and tested the resulting predictions against in vitro and clinical data (with collaborators). Based on these studies, a therapeutic T cell-based vaccine was designed, which is now being advanced to pre-clinical studies in monkeys. I will also describe work pertinent to how vaccination may induce broadly neutralizing antibodies that can neutralize diverse HIV strains â€“ this work lies at a crossroad of immunology, evolutionary biology, and statistical physics.

516082 Predicting Protein Interactions to Enhance Stability and Solubility of Therapeutic Antibodies

Abstract:

Therapeutic proteins are one of the fastest growing classes of innovator drug candidates. However, proteins are often problematic with respect to (mis)folding, aggregation, and practical limitations for solubility and solution viscosity at high protein concentrations. This presentation focuses on a combined experimental and computational approach to quantifying and predicting protein-protein interactions that can determine the physical stability of proteins such as therapeutic antibodies. Examples will highlight coarse-grained models to quantitatively or semi-quantitatively predict well-behaved vs. poorly behaved candidate molecules, spanning from antibody fragments to full-length monoclonal antibodies. Outstanding challenges will also be illustrated that underlie a number of common assumptions regarding how protein-protein interactions mitigate key pharmaceutical product attributes such as solution viscosity and aggregation rates.

527965 The Putative Liquid-Liquid Transition Is a Liquid-Liquid Transition in Some Atomistic Models of Water

Abstract:

Numerous computational studies by independent research groups have provided strong numerical evidence indicating that the ST2 water model exhibits a metastable liquid-liquid phase transition (LLPT) under deeply supercooled conditions. A similar LLPT has been hypothesized to occur in real water to provide a thermodynamically consistent explanation for its anomalous behaviors, such as its rapid increase in compressibility upon cooling in the supercooled regime. The existence of an LLPT in ST2 and other models of water has been challenged by Limmer and Chandler [J. Chem. Phys. 135, 134503 (2011); J. Chem. Phys. 138, 214504 (2013)], however, who observed signatures of only a single liquid phase in free energy calculations performed near ST2â€™s LLPT. These findings fueled vigorous debate over the phase behavior of supercooled ST2 water, which has been ongoing for nearly a decade. Although recent studies have provided unambiguous evidence to confirm the existence of a LLPT in ST2, the origin of the discrepancy with the Limmer-Chandler (LC) studies remains unresolved.

Here, we provide a resolution to this long-standing controversy by analyzing the hybrid Monte Carlo (HMC) code that Limmer and Chandler used to perform their free energy calculations for ST2. We show that the LC HMC code suffers from a fundamental design flaw that introduces significant sampling errors. These errors distort ST2â€™s equation of state and are easily detected using three rigorous statistical mechanics based consistency checks. When we correct the LC code, we recover the correct equation of state of ST2, in excellent agreement with two independent codes; the corrected code also satisfies the consistency checks. The corrected code gives clear evidence of two liquid phases in the free energy surface, while the uncorrected code reproduces the LC published results, which show a single liquid basin at conditions where several other studies have reported two liquid phases.

529079 Water-Mediated Interactions Involving Heterogeneous Molecules, Surfaces, and Phases

Abstract:

Hydrophobic interactions are a fundamental driving force for self-organization and self-assembly in biology. From a theoretical point of view, both hydrophobic solvation and interactions have been studied extensively for model solutes and interfaces. However, many molecules of interest are highly heterogeneous in both chemistry and conformations (e.g., through configurational fluctuations in non-rigid systems). In this talk, I discuss our recent work to understand the ways such heterogeneities can impact water-mediated interactions, leveraging collaborations with experimental groups using model systems that can be well-controlled and characterized. I also discuss the application of a variety of techniques â€“ including nonequilibrium simulations, optimization algorithms, and coarse-graining strategies â€“ that help illuminate the physical picture.

532574 A Density Functional Theory Approach to Electrocatalytic Reaction Barriers

Abstract:

The activity and selectivity of electrocatalysts is dictated by the activation barriers of elementary reaction steps. These elementary steps are often a combination of chemical and electrochemical steps, where electrochemical steps involve the transfer of ions from/to the electrolyte and electrons from/to the electrode. Though density functional theory methods have frequently been applied to investigate elementary reaction (free) energies within electrocatalytic reactions, the calculation of barriers for elementary electrochemical steps is less frequent. It is challenging to represent the electrochemical nature of these reactions, as transitioning from an ion in the bulk of the electrolyte to a reaction event at the surface challenges the length and time scales accessible with typical DFT models. In this talk, I will give an overview of methods currently in the literature to consider electrocatalytic reaction barriers, and review a simple and transferable approach we have developed and applied. Our approach uses DFT calculations on analogous non-electrochemical reactions (ie. hydrogenation) to define transition state structures and a reaction coordinate that can be used to determine electrode-potential dependent barriers for equivalent electrochemical reactions (ie. proton and electron transfer). The non-electrochemical transition state is referenced rigorously to the bulk phase ion and electron chemical potentials. Approaches within a Marcus Theory framework to determine the potential dependence of barriers will be discussed, with application to elementary steps involved in N_2 reduction, CO_2 reduction, and furfural oxidation.

526328 The Influence of Local Environment on Theoretical Calculations of Adsorption and Reaction for Catalyzed Reactions

Abstract:

Advances in algorithmic efficiency and computer speed have made it possible to carry out calculations of the thermodynamics and kinetics of elementary processes involved in reactions occurring on heterogeneous catalysts that now approach chemical accuracy. Work done in the last decade has revealed that the accuracy of computed results depends not only local interactions â€“ bonding and non-bonding â€“ but also the effects of the local environment. Examples of the latter effects include the influence of electrostatic and dispersive interactions of framework atoms in a zeolite that are distant from the active center and influence of solvation and direct effects of cations and anions with adsorbed species involved in electrochemical reactions. This talk will illustrate how such interactions of adsorbed species with the local environment affect the enthalpies and entropies of species adsorption and activation for elementary processes occurring in zeolites and at the surface of electrocatalysts.

528857 Tackling the Inverse Design Problem in Quantum Chemistry

Abstract:

Modern quantum chemistry facilitates unprecedented predictions of molecular and materials properties in complex systems. However, there is a disconnect between the fundamental building blocks of quantum mechanics and synthetic chemistry, which becomes evident when contrasting the languages of indistinguishable electrons used in quantum mechanics and distinguishable functional groups used in synthetic chemistry. While there are well-established quantum mechanical approaches to translate a molecular structure into a predicted molecular energy or property, the inverse mapping from desirable property into optimal molecular structure is not obvious. As a result, optimizing molecular structure with respect to a given observable such as an energy (e.g., a reaction barrier in catalysis) or property (e.g., the polarizability of a dielectric material) typically requires an extensive approach that includes either guessing candidate structures and manually checking their suitability empirically or time-consuming training of empirical models to predict suitable new candidates. This work takes initial steps in formulating an inverse mapping approach that determines, entirely from first principles, a molecular structure optimized for a desired observable. To this end, we introduce an approach that partitions the response of a molecular wave function to an external perturbation into contributions from molecular fragments. Our approach, which we term linear response for molecular interactions, or LR(MI), allows us to distinguish additive and cooperative molecular contributions to a target property and thus identifies optimal molecular fragments for bottom-up property design that can be directly interpreted by synthetic chemists. As a result, our work goes beyond the traditional structure-to-property paradigm and provides a direct route to property-to-structure prediction. Therefore, this inverse mapping approach informs the rational design of molecules with desirable properties directly from first principles, thus having the potential to be truly predictive.

529416 Computational Alchemy to Drive Searches for Catalysts through Materials Space

Abstract:

First principles Kohn-Sham density functional theory (KS-DFT) enables the search for novel heterogeneous catalysts, but accurate and reliable calculations demand significant amounts of computational time that restrict the scope of searches through materials space. In general, computational effort needed for screening typically scales linearly with the number of systems. Clearly, expansive searches for novel catalysts will require developments in computational methods with significantly better scaling while preserving universality and reliability compared to KS-DFT. Here, we demonstrate the potential and limitations of computational alchemy for rapid predictions of descriptors for heterogeneous catalysis. We will present data showing the performance of computational alchemy with the oxygen reduction reaction on oxides, nitrogen reduction on nitrides, and hydrogen evolution reactions on carbides. We will critically evaluate the strengths and limitations of computational alchemy as a general method for catalyst screening through materials space.

525303 Synergistic Application of XPS and DFT to Investigate Metal Oxide Surface Catalysis

Abstract:

Density Functional Theory (DFT), and experimental surface characterization techniques like X-ray Photoelectron Spectroscopy (XPS) are often used independently by computational and experimental researchers, respectively, to provide insights into metal oxide catalysed reaction mechanisms, pathways, and energetics [1, 2]. It is vital to develop novel approaches that integrate these surface sensitive experimental techniques and theoretical tools to gain better understanding of transition metal oxide (TMO) catalysed surface reactions. However, determining an appropriate Hubbard U-correction term[3] is a challenge for the DFT community[4] and identifying realistic reaction intermediates and their corresponding XPS shifts is a challenge for experimental researchers. In this work, using CuO as a model TMO, we demonstrate an integrated experimental and theoretical approach to (i) determine the optimum U value range for TMOs, and, (ii) identify adsorbate/intermediate species on the TMO surface (and their XPS shifts), in a synergistic manner. The O 1s spectra of the as-synthesized CuO 2D nonoleaves is measured using XPS and show the presence of 4 different peaks with core level binding energies (CLBEs) of 529.7 eV, 531.4 eV, 533.2 eV, and 534.6 eV. DFT+U is used to calculate the CLBE shifts for the possible adsorbed moieties (atomic and molecular oxygen, molecular and split water, carbon dioxide, and carbonate), in various configurations, on both, clean and vacancy defect containing surfaces. We scan the theoretical CLBEs over a range of U values and compare the experimental and theoretical CLBEs of the aforementioned adsorbates, to provide an appropriate U value that predicts the experimental shifts correctly. Among the structures, the experimental shift of (i) +1.7 eV falls only in the core-level shift range for O_2 in the Î·1(O) configuration (+0.56 to +1.87 eV), (ii) +3.5 eV falls only in the core-level shift range for the adsorbed H_2O at the surface oxygen vacancy site (+2.87 to +3.79 eV), and (iii) +4.9 eV falls only in the core-level shift range for the adsorbed HCO_2 (resembling the structure of HCO_3, +4.81 to +5.02 eV). In the process of U value determination, we parallelly uncover the existing surface moieties, their adsorption configurations and their XPS shifts. The U value between 4-5 eV correctly establishes the experimental XPS spectral peaks to the respective adsorbates and their geometries. This U value differs significantly from the bulk properties fitted U value of 7 eV and is in excellent agreement with our recent DFT+U benchmarking study[5] based on the heat of adsorption and Fourier Transform Infrared Spectroscopy measurements. The integrated approach elucidated in this article, results in the identification of adsorbates/intermediates (and their CLBEs) for the experimental XPS spectral analysis, in addition to appropriate U value determination for DFT calculations, concurrently. The method demonstrated in this paper can be extended to other TMOs, as well as to advanced surface characterization methods like temperature programmed XPS and in situ XPS studies to gain mechanistic insights into complex reactions and to guide the design and development of efficient and novel TMO based catalysts.

REFERENCES

1. Varghese, J.J., Q.T. Trinh, and S.H. Mushrif, *Insights into the synergistic role of metal-lattice oxygen site pairs in four-centered C-H bond activation of methane: the case of CuO.* Catalysis Science & Technology, 2016. **6**(11): p. 3984-3996.
2. Tao, F. and M. Salmeron, *In Situ Studies of Chemistry and Structure of Materials in Reactive Environments.* Science, 2011. **331**(6014): p. 171-174.
3. Capdevila-Cortada, M., Z. Åodziana, and N. LÃ³pez, *Performance of DFT+U Approaches in the Study of Catalytic Materials.* ACS Catalysis, 2016. **6**(12): p. 8370-8379.
4. Cohen, A.J., P. Mori-SÃ¡nchez, and W. Yang, *Challenges for Density Functional Theory.* Chemical Reviews, 2012. **112**(1): p. 289-320.
5. Bhola, K., et al., *Influence of Hubbard U Parameter in Simulating Adsorption and Reactivity on CuO: Combined Theoretical and Experimental Study.* The Journal of Physical Chemistry C, 2017. **121**(39): p. 21343-21353.

534618 Maximal Predictability Approach for Identifying the Right Descriptors for Electrocatalytic Reactions

Abstract:

Density Functional Theory (DFT) calculations are being routinely used to identify new material candidates that approach activity near fundamental limits imposed by thermodynamics or scaling relations. DFT calculations are associated with inherent uncertainty, which limits the ability to delineate materials (distinguishability) that possess high activity. Development of error-estimation capabilities in DFT has enabled uncertainty propagation through activity-prediction models. In this work, we demonstrate an approach to propagating uncertainty through thermodynamic activity models leading to a probability distribution of the computed activity and thereby its expectation value. A new metric, prediction efficiency, is defined, which provides a quantitative measure of the ability to distinguish activity of materials and can be used to identify the optimal descriptor(s) Î”G_{opt}. We demonstrate the framework for four important electrochemical reactions: hydrogen evolution, chlorine evolution, oxygen reduction and oxygen evolution. Future studies could utilize expected activity and prediction efficiency to significantly improve the prediction accuracy of highly active material candidates.

527994 Understanding Structure-Property Relationships in Catalysts By Using Cluster Expansions

Abstract:

Density functional theory (DFT) is widely used to predict the structures and properties of materials, but its direct applications to nanomaterials of experimentally relevant sizes can be prohibitively expensive. It has been demonstrated that this problem can be addressed through the generation of the cluster expansion models trained by DFT. Cluster expansions are generalized Ising models that account for many-body interactions, and they are capable of accurately (within 5 meV / atom compared to DFT) predicting the energies of millions of structures per minute. In this talk, two examples of using cluster expansions to better understand structure-property relationships are given. In the first example, we compare predicted Pt-Cu nanoparticle structures with experimental characterization, and demonstrate that the best agreement is achieved by constructing a novel cluster expansion for alloy nanoparticles of varying shape and size that explicitly includes adsorbates. In the second example we present a study of the hydrogen evolution reaction (HER) on transition metal phosphide surfaces. Cluster expansions are used to predict structures and energetics of adsorbed hydrogen as a function of temperature and applied potential, allowing us to determine the potential-dependent activities of different sites while fully accounting for interactions among adsorbed hydrogen atoms.

540558 The Effects of Ionic Correlation Andsurface Polarization on Electrostatic Inter-Colloid and Inter-Emulsioninteractions

Abstract:

Electrostatics interactions provide tunability for the properties of soft materials. The electrostatic interactions are sensitive to various tuning parameters including solvent permittivity, electrolyte valence, size and concentration, and interfaces. Despite long-term usage of electrostatics in controlling the soft matter properties, such as buffer system in biology, the understanding of electrostatics interactions is still limited. Here we discuss about our researches on the effects of surface polarization on the electrostatic interactions between ion-containing emulsions, and the effects of electrolyte concentration on the electrostatic interactions between colloids. The researches shed light on why classical electrostatic theories such as the Debyeâ€"HÃ¼ckel theory and even the non-linear Poisson Boltzmann theory break down in recent electrostatic mediated experiments.

540454 Probing Enzyme Catalyzed Hydrolysis of Cellulose in Ionic Liquids Using Enhanced Sampling Techniques

Abstract:

Ionic Liquids could transform the biofuel industry from an inefficient multi-step process requiring harsh organics and extreme processing conditions into a one-step, environmentally friendly process. They possess a variety of unique properties (e.g. negligible flammability, and vapor pressure), are hailed as "green-solvents", demonstrate wide chemical and thermal stability, and can be easily recycled.[1] Unfortunately, ILs are known to negatively affect the activity of cellulases. With the goal of rapid conversion of biomass to glucose, molecular level detail can assist in the rational design of enzymes and ILs. The work in this poster will demonstrate our approach to gaining molecular level insight of the Cel5A catalyzed hydrolysis of cellotetraose in an aqueous IL 1-butyl-3- methylimidazolium chloride [BMIM][Cl] solution.

The key challenge in modeling these systems is to find the balance needed between computational cost and accuracy. The hybrid quantum mechanics/molecular mechanics (QM/MM) approach is a method that can be used to probe these enzymatic reactions with the strength of QM accuracy, and the speed of MM forcefields. Still, the timescales needed to observe these phenomena occur on much larger scales than are computationally affordable. Enhanced sampling techniques can be used to circumvent these issues. For example in metadynamics[2], a powerful enhanced sampling algorithm, a time dependent bias is applied to a few coarse slow degrees of freedom to enhance phase space exploration by discouraging the system from visiting previous states. These methods have been used on a number of glycoside hydrolase catalyzed reactions in water[3,4] however, these methods have never been applied to study reactions in ILs. Our results indicate that the presence of IL in the active site plays an important role on the thermodynamics of the glycosylation reaction. A number of variants of metadynamics have been developed to further improve sampling, and speed up convergence. This poster presents our results and the challenges of applying the enhanced sampling technique well-tempered[5] metadynamics with adaptive gaussians[6] to resolve molecular level insight of these systems.

References

[1] J. Serra Moreno, S. Jeremias, A. Moretti, S. Panero, S. Passerini, B. Scrosati, G.B. Appetecchi. "Ionic liquid mixtures with tunable physiochemical properties", *Electrochimica Acta.*, **2015**, 599-608.

[2] A. Laio and M. Parrinello. "Escaping free energy minima", *Proc. Natl. Acad. Sci. USA,* **2002**, 99:12562–12566.

[3] L. J. Petersen, P. Reilly, A. Ardèvol, C. Rovira. "Mechanism of cellulose hydrolysis by inverting GH8 endoglucanases: A QM/MM metadynamics study" *J. Phys. Chem. B.*, **2009**. 113(20), 7331-7339.

[4] A. Ardèvol, J. Iglesias-Fernández, C. Rovira, X. Biarnés, A. Planas. "Catalytic itinerary in 1,3-1,4-β-glucanase unraveled by QM/MM metadynamics. Charge is not yet fully developed at the oxocarbenium ion-like transition state" *J. Am. Chem. So.*, **2011**. 133(50), 20301-20309.

[5] A. Barducci, G. Bussi, M. Parinello. "Well-Tempered Metadynamics: A Smoothly Converging and Tunable Free-Energy Method", *Phys. Rev. Lett*, **2008**, 100(2), 020603.

[6] D. Branduari, G. Bussi, M. Parrinello, "Metadynamics with Adaptive Gaussians", *J. Chem. Theory Comput.*, **2012**. 8,2247-2254

523446 Computational Study of Electrochemical Reduction of CO_2 on Transition Metal /p-Block Hybrid Nanocatalysts

Abstract:

Nanocatalysis, the study of catalysts for various homogeneous and heterogenous industrial applications is a vastly researched field these days. Factors such as depleting fossil fuels, increasing fuel costs and rising CO_2 emissions are some of the major reasons for the need to research on promising (efficient and sustainable) catalyst technologies. Electrochemical reduction of CO_2 using heterogenous catalysis is one of the solutions to convert CO_2 to value added hydrocarbon fuels which can later be used in various industrial applications like fuel cells. Developing a non-precious catalyst with high energy efficiency which can reduce the overpotential without compromising with the performance of the catalyst and selectivity of the products is a major challenge. Till date, Copper and its oxides are the only effective catalyst which can reduce CO_2 to hydrocarbons such as methane and methanol.

Our work mainly focuses on computational screening of different nanocatalysts for CO_2 electroreduction. In particular, we study transition metal/metal oxides â€“ p-block hybrids (non-precious metal catalysts) that can electrochemically convert CO_2 to value added fuels which can satisfy all or at least some the characteristics of an effective catalyst as discussed above. We use density functional theory, DFT calculations to find the adsorption properties of each intermediate species to study WO_3, MoO_3, ZnO and ZnS surfaces for CO_2 electroreduction by generating various possible reaction pathways from CO_2 to CH_4 and CH_3OH. These studies may not be enough to decide which is the best for CO_2 electroreduction, however, these will help in designing similar catalysts and can be a basis to do experimental study for the same process.

523818 From Metal to Plastic: Computer-Assisted Material Design for Marine Engine Non-Structural Components

Abstract:

Marine engine industry is in the quest for continuous improvement of efficiency and performance. Currently, all components involved in marine engine construction are made of metallic alloys. In order to reduce costs and weight, new materials for non-structural components must be identified. The introduction of new materials, e.g., nano-engineered thermoplastic polymers (NETP), will allow additional benefits due to drastic weight reduction and simplified maintenance and inspection operations. Advancements in NETP design and application in marine engine industry relies on computer-assisted multiscale material design (CAMMD). Indeed, by advanced CAMMD techniques, the structure of NEPT materials can be tailor-fitted to achieve the expected performances required by specific, advanced applications. Since the introduction of plastic materials in the construction of non-structural components for marine engines constitutes an element of great innovation, a specific rule framework must be defined yet. In this work, starting from the analysis of the regulatory context currently used for metallic alloys a certification procedure is proposed and applied to a case study related to the cylinder head cover of a four-stroke marine engine. In particular, the mechanical properties of a new NETP material designed by CAMMD have been verified trough a finite element simulation carried out on the relevant model.

523828 Stay Double, Stay Homologous: Combined Computational/Esperimental Approaches to RAD51/ssDNA Interactions in DNA Damage Repair

Abstract:

DNA replication, repair, and recombination proteins form complex and agile networks. These networks organize the participating proteins into molecular machines that act on different substrates and channel them to different outcomes. Some of these machines display the capacity to accurately repair DNA damage or reestablish damage DNA replication forks without the loss of genetic information. Under other circumstances, action of the same molecular machines destabilizes the genome, leading to cancer or to the accumulation of toxic repair intermediates resulting in cell death. The central player in homologous recombination (HR) is the RAD51 DNA strand exchange protein (aka recombinase); its inactive conformation is tightly controlled by a double-heptameric doughnut-like assembly while its active form is a nucleoprotein filament assembled on the single-strand DNA generated at the site of DNA damage. In order to get a comprehensive structural and thermodynamic perspective of RAD51/ssDNA interaction, in this work we will present an application of large scale, parallel atomistic and coarse-grained (CG) simulations, coupled with high resolution imaging techniques and biophysical methods (ITC and CD) to i) dissect the docking mode of RAD51 to ssDNA, ii) deriving the energetics of the interface in the protein/nucleic acid interface, iii) uncovering the role of DNA and protein flexibility in their assembly formation, and iv) determining the mechanism of RAD51 binding to ssDNA.

524146 Influence of Basis Set on the Electronic Structure and Physico-Chemical Properties of the Cerium Tribromide and the Cerium Tricloride: Two Lanthanide Compounds

Abstract:

The electronic structure and physico-chemical properties of the two lanthanide compounds known as cerium tribromide ($CeBr_3$) and cerium trichloride ($CeCl_3$), was studied in this work, using the Unrestricted ab initio Hartree-Fock (UHF) and density functional theory methods (DFT) (B3P86, B3PW91, B1LYP, X3LYP and B3LYP) with the assistance of GAUSSIAN 09 software. Different basis set such as CEP-4G; CEP-31G and CEP-121G were used to determined the dipole moment, the atomic charges, the polarizabilities tensors, the average polarizability, anisotropy, the first molecular hyper-polarizability, the electric field, the HOMO- LOMO energy gap. The optoelectronic and some thermodynamic properties of the title molecules were also determine and interpreted. From our results we found that, basis set can modify the electronic structure and the physico-chemical properties of molecules. The values of Egap show us that these compounds can be considered as superconductors and be use to increase the storage capacities of the computers and make them faster.

524795 Multi-Scale Simulations of Biomacromolecules for Design of Biomaterials

Abstract:

In this poster we will present our groupâ€™s multi-scale simulation studies of biomacromolecules for design of biomaterials. In the first part of the poster we will present our ongoing work using a combination of atomistic and coarse-grained molecular dynamics simulations to understand phase-transitions in soft materials made of thermoresponsive biopolymers, specifically elastin-like peptide (ELP) and collagen-mimicking peptide (CMP).[1, 2] In the second part of the poster, we will present our ongoing work using simulations with atomistic and MARTINI-type coarse-grained models to predict structure within synthetic melanin nanoparticles via assembly of oligomers of 5,6-dihydroxyindole (DHI) and 5,6-dihydroxyindole-2-carboxylic acid (DHICA).

1. E. Condon[#], T. B. Martin[#], A. Jayaraman* "Effect of Conjugation on Phase Transitions in Thermoresponsive Polymers: An Atomistic and Coarse-Grained Simulation Study" *Soft Matter* (2017), 13, 2907 ([#] equal authorship)
2. E. Condon, A. Jayaraman*, "Development of Coarse-Grained Model of Collagen-like Peptide (CLP) for Studies of CLP Triple Helix Melting " *J. Physical Chemistry B*. 122 (6), pp 1929â€“1939 (2018)

526808 Binding Rates of Polyaromatic Hydrocarbons during Soot Formation: Insights from Reactive Molecular Dynamics

Abstract:

In gas-phase synthesis of soot nanoparticles, polyaromatic hydrocarbons (PAHs) form and grow by collisions at high temperatures. While a hard sphere equation is frequently used to predict the sticking rate coefficient, this equation neglects the influences of potential interactions and internal energy of the colliding entities.

Here, a collision rate theory-Molecular Dynamics simulation approach is used to calculate sticking probabilities and dimerization rate coefficients of PAH molecules. The probability of dimerization depends upon the relative velocity between the two colliding PAH clusters and their initial impact parameter and orientation. For all PAH molecules investigated here (naphthalene, pyrene, pentacene and coronene) there is a well-defined region of relative velocities and impact parameters where sticking is highly probable, while outside this region probability drops to zero. The dimerization rate of such PAH molecules is determined by quantifying their binding probability at various process temperatures. Integration over all impact parameters and velocities for collisions (in accordance with the Maxwell Boltzmann distribution) yields the dimerization rate coefficient for a given PAH structure. Criteria for successful dimerization events are proposed based on their collision time.

The accurate binding rates obtained by the presented method can be applied in developing rate equations to predict material formation and growth rates in vapor phase systems.

526829 Capturing Non-Ideal Surfactant/Nanoparticle Interfacial Structure with Variable Coverage Molecular Simulations

Abstract:

Surfactants adsorbed to hard nanoparticle interfaces are poorly understood despite their importance in nanotechnologies because of their complex non-ideal behavior, the non-deformable nanoparticle interface, and the similar length scales of surfactant molecules and surfaces. We develop a molecular thermodynamic approach to directly simulate isotherms of surfactants at interfaces with arbitrary coverages and to describe surfactants that are not well described with simple molecular thermodynamic theories. All-atom molecular simulations are used to estimate the chemical potential of non-ideal surfactant structures by calculating the free energy required to concentrate a fixed number of adsorbed surfactants at low density and under ideal conditions onto a small subsection of the interface with high local density. Standard adsorption calculations are used to reference the low density state to the bulk liquid surfactant potential, thus providing equilibrium coverage estimates. This approach captures non-ideal effects such as specific surfactant-surfactant interactions, amphiphilic surfactant structure, and more than one monolayer of adsorbates. We show that this new approach reproduces analytical results for hard spheres at the interface and can resolve qualitative changes in surfactant structure at between low and high loadings.

527418 Challenges and Strategies of Modeling Extra-Framework Metal Cations in Zeolites from First-Principles: Knowledge Learned from Cationic Iron Exchanged in SSZ-13

Abstract:

Metal-exchanged zeolites are widely used as catalysts for various practical applications. The extra-framework cationic metal centers dispersed throughout zeolite support are typically the active sites responsible for the catalysis. Quantum chemistry methods, e.g. density functional theory (DFT), become powerful as to computationally model these cations for rationalization of the corresponding catalytic functions. Accurate modeling, however, involves different challenges, ranging from the reliability of DFT in capturing correct electronic structures of metal cations and relevant properties, to the heterogeneity of the anionic support that charge-compensate metal cations, e.g. framework aluminum distribution and proximity, to the complexity of cation speciation locally, etc. In this work, we use our study on modeling cationic iron (Fe) siting and speciation in SSZ-13 zeolite as an example, to illustrate how these unavoidable challenges can lead to incomplete or even erroneous conclusions if not properly handled, and come up with corresponding strategies. Through benchmarking with wavefunction coupled-cluster method, we parameterize DFT functionals that are more reliable in predicting cationic Fe properties, e.g. coordination preference. A geometry enumeration algorithm is developed to ensure effective (avoiding redundant computational effort) and complete encompassing of Al proximity effect on metal cation siting. Local cation speciation is predicted through comprehensive sampling of cation species followed by ab inito thermodynamics analysis. Integration of these steps allows more systematic and accurate modeling of cationic Fe siting and speciation in SSZ-13, and the strategies are readily transferable to zeolitic metal cation modeling in general within any zeolite topologies of interest.

528158 Molecular Simulations of Liquid-like Assemblies of Intrinsically Disordered Proteins

Abstract:

Proteins, which undergo liquid-liquid phase separation (LLPS) to form dynamic assemblies within biological systems, have been gaining much attention due to their relevance to membraneless organelles and prospects for material design. Much work has been directed toward understanding the molecular mechanisms of phase behavior and predicting the phase boundaries and dynamics of biomolecules within these assemblies from the composing protein sequences. Due to the scale of these systems, atomistic simulations would be severely limited in scope, thus motivating the development of appropriate coarse-grained models. Using coarse-grained molecular dynamics simulations, we are able to consistently model condensed and dilute phase properties of disordered proteins, and show that many of these sequences follow the behavior of homopolymers. Using these relations, we can predict phase behavior of protein sequences using simulations of a single chain.

528355 Competitive Adsorption of Toxic Gases in a Humid Environment: Insights from Density Functional Theory

Abstract:

Toxic gases such as ammonia, phosphine, and methyl bromide, are commonly used in industry and agriculture, which places workers at risk of accidental exposure. Most widely used filtration methods rely on activated carbon, which has low capacity and selectivity for these gases. However, in recent years, researchers have looked to other porous materials, such as metal-organic frameworks (MOFs), as sorbents for toxic gases. [1] One benefit of using MOFs is they can be tailored to specific adsorption applications by adding functional groups or metal sites which increase the MOFâ€™s affinity for the target molecules. [2]

We used density functional theory to study interactions of ammonia, phosphine, arsine, and methyl bromide with 33 metal catecholates in order to examine periodic trends in the binding strength between the gases and metals. [3]

We find markedly different behavior in ammonia than in phosphine and arsine due to different processes governing the adsorption despite the apparent similarity of these molecules. Phosphine and arsine adsorption is primarily driven by electron density being donated from the adsorbate molecule to the metal atom, while ammonia binding involves both some electron donation and a significant Coulombic attraction between the negatively-charged N atom and the positively-charged metal atom. We also studied simultaneous adsorption of methyl bromide and water on metal catecholates. While late transition metals such as Pt and Au have the highest affinity and selectivity for MeBr, we find that alkaline earth metals such as Ca are more resistant to the presence of humidity.

We will discuss how these different adsorption mechanisms are manifested through trends in binding strength across the periodic table, as well as the implications for adsorption of industrially relevant toxic gases in humid environments.

1. S. Bobbitt, M. L. Mendonca, A. J. Howarth, T. Islamoglu, J. T. Hupp, O. K. Farha, R. Q. Snurr, *Chemical Society Reviews,* 46, 3357-3385, 2017
2. Yu, P.Ghosh, and R. Q. Snurr,*Dalton Transactions*, 41, 3962-3973, 2012
3. S. Bobbitt, R. Q. Snurr, *Ind. & Eng. Chem. Res*., 56, 14324-14336, 2017

528483 CO_2 Adsorption in Nickel Based Metal Organic Framework Ni-DABCO: A Density Functional Theory and Grand Canonical Monte Carlo Study

Abstract:

The continuous emissions of green house gases, especially carbon dioxide, are still generating concerns over the adverse effects they are having on the planet. Metal organic frameworks (MOF) have obtained wide spread attention for their CO_2 adsorbing capabilities. MOFs are formed through the coordination of inorganic metal clusters with organic bridging ligands. Ni-DABCO is one of the many MOFs that are being studied for CO_2 separation. It is part of the M-DABCO MOF series, which has isostructural zinc, copper and cobalt variants. The organic linkers in M-DABCO are 1,4-benzenedicarboxylate (BDC) and 1,4-diazabicyclo[2.2.2]octane (DABCO). The latter is also known as triethylenediamine (TED). We have computational studied the interaction of CO_2 in Ni-DABCO using periodic density functional theory (DFT) and gran canonical Monte Carlo (GCMC). Our periodic DFT calculations demonstrate that upon adsorption, CO_2 induces small changes on the unit cell parameters of the Ni-DABCO framework. Our calculations also demonstrate that the electrostatic potential of the framework slightly changes as the number of CO_2 molecules increased in the system, suggesting that upon CO_2 adsorption the electron densities also change in the framework. The partial atomic charges of the framework atoms close to the CO_2 molecules, as well as the CO_2 molecules exhibited a change in their atomic charges when compared to the isolated systems. GCMC simulations were used to generate CO_2 adsorption isotherms on Ni-DABCO at 298K, and our results were compared with experimental isotherms reported in the literature. Furthermore, the textural properties of the material, including BET area and pore volume, were calculated using a simulated N_2 isotherm at 77K and were also compared to published results. Our simulated isotherms and the simulated powder X-ray diffraction patterns are in good agreement with experimental results. However, at pressures higher than 10 atm, the simulated isotherm underestimates the experimental isotherm, suggesting that this material may expand during adsorption and could be a pressure dependent phenomenon.

529048 A Comparison of Crystalline and Icosahedral Order inÂ Ag_6Cu_4Â and CuAu Alloys

Abstract:

By utilizing molecular dynamics, we can compare crystalline and icosahedral order during crystal growth in Ag_6Cu_4 and CuAu alloys. As temperature is decreased, the Ag_6Cu_4 alloy growth rate decreases, whereas the CuAu alloy shows a growth rate increase. With that being said, the nucleation process in the Ag_6Cu_4 alloy is immediate when on a Pt interface but when on a Cu interface, the process exhibits a slower rate. The crystallization process for the CuAu alloy on a Pt interface occurs over a longer period of time than that of the Ag_6Cu_4 alloy, but in the vicinity of a Cu interface, the time interval is considerably slow. By changing the amount of supercooling and varying the interface, we are able to examine the close-packed (CP) and icosahedral (Ih) order exhibited and understand how these changes impact the system.

529423 Molecular Dynamics Simulation of Modified Nafion 117 Based Anion Exchange Membrane Fuel Cell: Transport and Nanophase-Segregated Structure Properties

Abstract:

In this research, we investigate the effect of structural variation of Nafion 117 on the nanophase-segregation and transport properties using molecular modeling approach. In this study we replace the sulfate group of nafion117 with two different phosphate groups, and characterize its effect on nanophase-segregated structure and transport of water/protons at 80°C with various water contents. The two groups that will be tested are: changing the entire tail group to a polyvinyl phosphate, and changing the tail group to a difluoromethandiphosphonic acid. This will be done by simulating the new molecular structures using molecular dynamic simulations at 80°C under isobaric conditions using LAMMPS. Dreiding force field will be partially modified to handle perfluorocarbon compounds, water, and hydronium. The equations of motion will be integrated using Verlet algorithm with a 1-fs time step, and the PPPM method to evaluate the electrostatic interactions. After an equilibration cycle, each system will then be simulated for an extra 2 ns to sample structures for statistical treatment. By simulating the proton conductivity of these molecules we seek to understand the relationship among underlying polar groups, nanophase-segregated structure, and transport which would serve in designing new polymer electrolyte membranes.

529493 Efficient Generation of Polymer Amorphous Structure By Reverse-Mapping from Beads-Spring to Full-Atomistic Model

Abstract:

Structure generation is the beginning in investigating materials with molecular dynamics simulation, however, we have sometimes difficulties in generating geometries of molecular systems, e.g. polymer amorphous structures, micro phase separation structures, etc. Since polymer systems have long relaxation time, it is often difficult to equilibrate the systems. There are several methods for polymer amorphous structure generation utilizing Monte-Carlo (MC) based simulations such as Rotational Isomeric State (RIS)-MC and Configurational Bias (CB) MC. Another method is to consider a relation between real polymers and universal polymer models. A mapping relation between beads-spring (BS) model and full-atomistic (FA) model is proposed by Evaraers group [1]. According to the mapping relation, we have developed a scheme which constructs a FA polymer model from BS model utilizing reverse-mapping technique. In our scheme, BS model, coarse-grained model, and FA model are successively constructed. Our scheme provides an efficient and easy way to construct FA polymer amorphous structures.

[1] C.Svaneborg et al., "Kermer-Grest models for universal properties of specific common polymer species", arXiv:1606.05008

530442 Structural and Vibrational Properties of a Si- and Se-Induced 216-Atom Quasi-Random Ingaas

Abstract:

We are using a novel computational approach to study the structural and vibrational properties of Si and Se dopants and vacancies in randomly ordered 216-atom simulation boxes to represent $In_{0.5}Ga_{0.5}As$ based on density function theory (DFT). Specifically, the random structure is modeled atomistically using a newly implemented efficient search and optimize algorithm. We want to compare and contrast the effect of Si and Se dopants in the vicinity of a vacancy to construct a model for a desirable *n*-type InGaAs. Different local environments for each case are determined using a modified cell list algorithm that I wrote. Vibrational frequencies are obtained from dynamic matrices using a Green's function. And the dynamic matrices are calculated from weighted averages of local environments procured through two processes: a pure InGaAs system studied through the force field fitting method and a subsystem within the defect strain field investigated through the so-called "frozen phonon" method.

There are two unique aspects behind this project: the construction of a large quasi-random structure and obtaining vibrational frequencies through a "frozen phonon" approach. In the past, most researches have only studied ordered structures since it is surprisingly difficult to produce a truly random structure especially for a large supercell. A quasi-random structure considering multi-body clusters has never been constructed on such a large scale. Our approach combined an effective search of the multi-body (including two-body to four-body clusters) nearest neighbor (up to sixth nearest neighbor) with optimizing with a multi-objective simulated annealing to successfully generate a quasi-random system, that could contain thousands of atoms, at a very small computational cost. The second aspect that makes my research unique is the implementation of the frozen phonon approach. Due to the relatively large size of the system when compared to normal *ab initio* simulation systems, we cannot apply the phonon mode calculations in the existing modeling package. So I used an alternative way to solve this problem by applying the frozen phonon method for all the atoms within the dopant's and/ or defects' strain field to obtain the dynamic matrices and then further convert them to vibrational frequencies through a Green's function.

530625 Modeling Rosette Nanotubes Using the Martini Forcefield

Abstract:

Modeling Rosette Nanotubes Using the MARTINI Forcefield

Vyshnavi S. Karra, Hicham Fenniri, Francisco R. Hung

Department of Chemical Engineering, Northeastern University

Rosette nanotubes (RNTs) are biocompatible supramolecular nanostructures that are formed via self-assembly of building blocks of Watson-Crick DNA-inspired guanine-cytosine (G C) motifs. Similar to the double helix of DNA, hydrogen bonding between the individual motifs lead them to assemble into rings, called rosettes.1,2 A combination of π-π interactions between the rings and hydrophobic effects lead the rosettes to self-assemble into nanotubes. There are 2 types of RNTs, depending on how the nanotube is self-assembled: either the rings are stacked or they are assembled into helical coils. Because of its biocompatibility, RNTs have attracted attention for drug delivery and biological applications, such as encapsulating dexamethasone to enhance cell growth in bones.3 However, a fundamental understanding of the interactions of RNTs with cell membranes, proteins and other biomolecules could lead to the development of optimal RNTs for drug delivery and other applications. For example, nanotubes tend to "stick" to the blood vessel walls at higher probabilities than spherical nanoparticles, making them more difficult to implement as drug carriers.4 Using the MARTINI force-field,5 we are developing a coarse-grained model of these RNTs. Here, we present details of our model of the G C motifs, and report classical molecular dynamics simulations of individual motifs, rosettes and small RNTs in polarizable water. There are 2 main objectives of these MARTINI simulations: 1) to fine-tune the interactions between the different species in our systems, and 2) to assess the stability of the assembled structures. Results will be compared against existing atomistic simulations and experimental data.

1. H. Fenniri, B. L. Deng, and A. E. Ribbe. J. Am. Chem. Soc. 2002, 124, 11064-11072.
2. H. Fenniri, B. L. Deng, A. E. Ribbe, K. Hallenga, J. Jacob, and P. Thiyagarajan, Proc. Natl. Acad. Sci. USA 2002, 99, 6487-6492.
3. G. Borzsonyi, R. Johnson, A. Myles, J. Cho, T. Yamazaki, R. Beingessner, A. Kovalenko, and H. Fenniri. Chem. Commun., 2010, 46, 6527-6529.
4. C. Kinnear, T. Moore, L. Rodriguez-Lorenzo, B. Rothen-Rutishauser, and A. Petri-Fink. Chem. Rev. 2017, 117, 11476-11521.
5. J. Uusitalo, H. Ingólfsson, P. Akhshi, D. Tieleman, and S. Marrink. J. Chem. Theory Comput., 2015, 11, 3932–3945.

530698 Conformal Sites Model for Adsorbed Films on Energetically Heterogeneous Surfaces

Abstract:

Molecular simulations and theoretical analyses of thin films adsorbed onto solid surfaces or within narrow pores have usually treated the solid surfaces as atomically smooth and energetically homogeneous. Most surfaces, however, exhibit roughness through geometric curvature, variations of surface composition and chemical groups attached to the surfaces. Such oversimplification of the surfaces thus misses important features of adsorption, especially at low bulk pressure. In this research, a flexible conformal sites model, mapping a real interfacial system with an energetically heterogeneous surface onto a reference system with an energetically homogeneous surface, is developed. Grand canonical Monte Carlo simulation is used to explore the limits of applicability of this model to the prediction of adsorption isotherms. Preliminary simulation results show that our conformal sites theory is quite successful in reproducing the adsorption behavior in non-polar systems and in slightly heterogeneous polar systems. Due to the simplicity and accuracy of the theory, we anticipate this new method may be incorporated in the software of experimental instrument to study porous materials by taking advantage of the corresponding states theory.

530848 Screening of Bio-Based Plasticizers for Poly(vinyl chloride) and Poly(lactic acid) Via Atomistic Simulations

Abstract:

There is a growing concern for the toxicity of industrial polymer plasticizers to humans. Especially, phthalate derivatives are very common plasticizers and exhibit toxic effects by degradation. These plasticizers are particularly widely used in the processing of polyvinyl chloride (PVC). On the other hand, bio-based polyesters have attracted much attention due to their biodegradability and biocompatibility that offer clear advantages for both customers and environment. Polylactic acid (PLA) is an attractive candidate for replacing petrochemical polymers because it is from renewable resources. However, PLA needs to improve some properties by an inclusion of additives to use widely in industrial scale.

In our study, various bio-based molecules with potential to replace phthalate derivative plasticizers for PVC and PLA have been modeled by employing atomistic scale simulations. Our approach depends on molecular dynamic simulations to determine polymer/plasticizer interactions and observe plasticization effects via the lowering of glass transition temperature (Tg). Structural stability was evulated by solubility behavior of plasticizers in polymer and to describe Tg values and volumetric behavior of polymer/plasticizer mixtures, pressure-volume-temperature (PVT) relationship was used. Selected bio-based molecules including levulinic acid esters, succinic acid esters, oleate esters, dioctyl adipate and for comparison one of the phthalate derivative and acetyl tributyl citrate were constructed with TRAPPE-UA force field, were optimized and their densities similar to literature were obtained. Then, the atactic PVC homopolymer chains consisting of 150 monomers with equal fractions of meso and racemic diads distributed randomly along the chain were constructed based on the Generalized Amber Force Field (GAFF). Amorphous PLA model was constructed as PLLA consisting of 150 monomers with PLAFF3 force field. Tg and densities of pure PVC and PLA were determined as 344 K and 1.38±0.01 g/cm3 for PVC, 357 K and 1.211±0.01 g/cm3 for PLA. Polymer/plasticizer structures contained 10% to 30% plasticizer by weight to predict the performance of the plasticizers. Bio-based derivative esters, including levulinic acid esters and succinic acid, were also constructed with different substituents in the ester groups to observe chain length effect on plasticization.

The variation of glass transition temperatures the polymer was investigated and compared as a function of plasticizer type and content by simulating PVT relationship and stability of the polymer/plasticizer mixtures were compared by calculated solubility values. The compatibility of plasticizers was calculated through the cohesive energy densities of the systems and Flory Huggins interaction parameters were obtained by interpolation from energy-volume fraction plots.

This work was supported by the Scientific and Technological Research Council of Turkey (Grant# 114M178) and the Scientific Projects Unit of Istanbul Technical University (Grant# MGA-2017-40714).

530895 A Combined Molecular Dynamics and Experimental Study of an Imidazolium Based Ionic Liquid Electrolyte Solution for Low Temperature Applications

Abstract:

Imidazolium based ionic liquids are well known for their versatility as a solvent for room temperature and high temperature applications; however, their low temperature characteristics have not been studied. This investigation utilizes computational and experimental work to further study and design an imidazolium based ionic liquid electrolyte solution for space applications that can retain liquid properties at extremely low temperatures. The solution, containing an imidazolium based ionic liquid, water, lithium salt, and ionic liquid co-solvent, was modeled using the Gromacs 5.1.4 package in a temperature range of 25 °C to -120 °C. For each simulation, the molecules were randomly inserted into a 5x5x5 nm box and simulated under an NPT ensemble. Radial distribution functions indicate increased interactions and possible hydrogen bonding between water molecules and anions as temperature and water content is decreased. Self-diffusion coefficients were also calculated and show a nonlinear decline as the solution reaches colder temperatures. Experimental data including differential scanning calorimetry curves, viscosity, and conductivity measurements further support the trends from the simulations. Water concentration has a strong effect on glass transition temperature, viscosity and conductivity of the studied system. The results from this investigation show potential for ionic liquid electrolyte systems to be employed in low temperature applications.

531122 Biasing High-Dimensional Free-Energy Landscapes for the Detection of Stable Clusters in Self-Assembling Systems

Abstract:

The self-assembly of proteins, and nanoparticles is a complex phenomenon driven by multi-scale, hierarchical interactions with multiple association pathways. While molecular simulations of these systems provide key insights into these processes, high energy barriers of association pathways can trap the simulation in metastable states and impede comprehensive exploration of assembling structures.

In this poster, I will highlight how recently introduced enhanced sampling techniques, which allow users to bias multiple degrees of freedom (or collective variables), make simulations of self-assembling systems tractable.[1,2] This method shows rapid convergence of free energies for 3-13 particle Lennard-Jones systems. The simulation trajectories are post-processed to detect stable structural minima and find the relative stability of structures. We apply the same strategy to detect the structure of stable water clusters and amyloid-beta peptide aggregates. In the course of studying these systems, we also comment on the best choice of collective variables for such high-dimensional, self-assembling systems.

REFERENCES

(1) Pfaendtner, J.; Bonomi, M. Efficient Sampling of High-Dimensional Free-Energy Landscapes with Parallel Bias Metadynamics. *J. Chem. Theory Comput.* **2015**, *11* (11), 5062â€"5067.

(2) Prakash, A.; Fu, C. D.; Bonomi, M.; Pfaendtner, J. Biasing Smarter, Not Harder, By Partitioning Collective Variables Into Families. (*Under Review*). **2018**.

532330 GPU Accelerated Experiment Directed Metadynamics for Scattering Profile Biasing

Abstract:

Molecular dynamics (MD) simulations can reveal key information about the behavior of materials at the atomic scale, but mismatches between experimental conditions and inaccuracies in force-fields can make comparison and iterative improvement from experimental data difficult. We can mitigate this issue by minimally bias collective variables to match ensemble averages through maximum entropy biasing techniques. With Experiment Directed Simulation, we can even match entire PMFs to experimental data. Examples of when a reference PMF is available include radial distribution function matching in coarse-graining or structure factors derived from small-angle x-ray scattering data. A major challenge is that these techniques require accurate evaluation of the structure factor, and its derivative, at each timestep. In this work, I will show our GPU-accelerated implementation of structure factor calculations, including treating background subtraction from reference simulated systems. This new approach enables both rapid evaluation of Monte Carlo searches of crystal structures and MD of solutions.

532948 Aggregation Behavior of 1-n-Dodecyl-3-Methylimidazolium Octylsulfate Biamphiphilic Ionic Liquid in Aqueous Solution

Abstract:

The structure and organization in an aqueous solution of ionic liquids for which either the cation or anion is responsible for inducing nonpolar domains are well understood. However, information regarding the aggregation behavior of such ionic liquids in aqueous solution is primitive at this point. The present contribution is aimed at addressing this question and focuses on a molecular dynamics study to probe aggregation features of aqueous 1-n-dodecyl-3-methylimidazolium [C_{12}mim] octylsulfate [C_8SO_4] solution and its vapor-liquid interface at room temperature. Previously, we have demonstrated that the presence of layer-like (polarâ€"nonpolarâ€"polar) morphology in bulk liquid phase of [C_{12}mim][C_8SO_4]. Our results indicate the formation of a single aggregate containing cations and anions of ionic liquids with evidence of lamellar morphology. The penetration of water in the nonpolar domain constituted by the alkyl tails is minimal, while the polar head groups are exposed to the bulk water. The introduction of vacuum interface induces spontaneous dissociation of the aggregate eventually leading to interfacial adsorption of cations and anions. Similar to surfactant molecules, the cationic and anionic polar heads orient towards the bulk water, while the nonpolar tails protrude out into vacuum. The structural characterization will be discussed in terms of number density profiles, orientation of the ions and aggregate number.

533008 Composition Effect on the Nucleation Process in CuNi Systems

Abstract:

By using molecular dynamics, we show the effect of the composition on the nucleation process. To do so, we use three different systems, pure copper, pure nickel and the copper-nickel alloy. The system is sampled with the umbrella sampling that allows us to look at its free energy. Our results show a higher free energy for the alloy than for the pure systems. Meanwhile, the alloy tends to form a bigger critical nucleus than the two other systems. The structure as well as the composition of the nucleus has also been studied showing how it has been impacted by the composition.

533897 Predicting Point Defect Concentrations in Complex, Disordered Oxides

Abstract:

Materials-by-design in rapidly becoming a standard in materials science, driven by substantial advances in computational capability and improved integration of theory and experiment. In order to continue to advance the materials design and discover, computation must continue to drive towards experimental reality through incorporation of finite temperature effects at relevant operating and/or synthesis conditions. Chemically and structurally complex solid compounds, including those with magnetic and atomic disorder, are rapidly extending new material functionalities and device performance across applications ranging from ceramic fuel cells and thermochemical redox cycles to membrane separations and sensors. Accelerated development of these complex compounds requires accurate predictions of material properties including effective defect formation energies and equilibrium defect concentrations at finite temperatures. Traditional first principles approaches employ periodic models and systematically examine relatively ordered atomic structures to identify the lowest energy defect sites are rarely suitable for describing the disorder and combinatorial scope of complex compounds. In this work we demonstrate a new strategy that incorporates atomic and magnetic disorder, ensemble descriptions of defect sites and formation energies, and effects beyond the dilute defect limit to predict the temperature dependence of oxygen vacancy concentrations in complex perovskite oxides, including $Ba_{0.5}Sr_{0.5}Fe_{0.8}Zn_{0.2}O_3$ (BSFZ), $Ba_{0.5}Sr_{0.5}Co_{0.8}Fe_{0.2}O_3$ (BSCF), and $BaCo_{1-x-y-z}Fe_xZr_yY_zO_3$ (BCFZY). We compare our thermodynamic predictions against experimental data to validate our methods.

533952 Reweighting Molecular Simulation Configurations for Rapid Force Field Parameterization

Abstract:

Molecular simulation is an efficient approach for predicting thermophysical properties, such as vapor-liquid equilibria (VLE), pressure-density-temperature behavior (*pÏT*), and caloric properties. The reliability of molecular simulation results depends primarily on the force field, with the non-bonded interactions playing a key role. Unfortunately, non-bonded potentials that are empirically parameterized with thermophysical properties (typically VLE) traditionally use iterative brute-force methods that require large amounts of molecular simulation.

A more efficient approach than simulating each proposed non-bonded parameter set is to reuse the information gained from nearby parameter sets. This study demonstrates how to utilize Multistate Bennett Acceptance Ratio (MBAR) for this purpose. Specifically, MBAR reweights configurations that are sampled using a few reference parameter sets to predict the internal energy and pressure for other non-bonded parameter sets, without performing additional direct simulations.

We demonstrate how to implement MBAR with NVT isothermal isochoric integration (ITIC) and Grand Canonical Monte Carlo (GCMC) to obtain vapor-liquid equilibria properties. The MBAR-ITIC and MBAR-GCMC algorithms require that the energies and forces are recomputed for the sampled configurations using the new non-bonded parameter set. Basis functions are shown to be a computationally efficient method for recomputing these quantities. Basis functions eliminate the need for looping through non-bonded interactions when a linear relationship exists between the non-bonded energy (or force) and the non-bonded parameters. Implementing basis functions with MBAR-ITIC and MBAR-GCMC reduces the computational cost to parameterize the non-bonded potential by approximately three to five orders of magnitude relative to the traditional approach, where the reduction is greatest for larger molecules and systems.

534074 Importance of Molecular Conformations on Dipole Moment and Thermophysical Properties Estimation

Abstract:

The knowledge of thermophysical properties, such as vapor viscosity and thermal conductivity, for a wide range of molecules is crucial for industrial applications. However, such properties are often not available for many species of interest, especially for proprietary compounds. Experimental measurement is expensive and time-consuming. Thermophysical property estimation methods are available that depend in part on the value of the gas phase dipole moment, which is difficult to measure experimentally. Alternatively, it has been shown that quantum chemical methods such as density functional theory (DFT) can be used to accurately estimate the dipole moment vector for simple molecules. However, dipole moments for molecules having a large number of conformational degrees of freedom have been very challenging to calculate. This is because the conformation of the molecule can greatly affect its dipole moment. In this work, we present an efficient computational formalism for estimating the gas phase dipole moment of molecules with many conformational degrees of freedom. We compare thermophysical properties estimated from dipole moments determined from experiments and from computations to establish the reliability of our approach.

534084 Computer-Aided Description of Materials Stability at the Nanoscale

Abstract:

Nanomaterials (e.g. alloyed metal nanoparticles (MNPs)) find diverse applications in fields from catalysis to medicine where their unique properties vastly differing from bulk (e.g. high catalytic activity, distinct bandgaps) are highly desirable. The properties of nanomaterials are largely determined by their structure [1-2], but understanding of the nanomaterials stability with structure variation (i.e morphology and chemical ordering) remains limited. The challenges encountered in deciphering structure-dependent thermodynamic stability in nanomaterials are i) the intractable number of potential nanomaterial morphologies and chemical ordering within each unique structure (e.g. in bimetallic MNPs) and ii) the presence of ligands that bind the nanomaterials surface to stabilize them at the nanoscale. Herein, we highlight our recent examples on gaining physical insights and accurately describing the stability of nanomaterials addressing both the aforementioned challenges, through first-principles-based computational modeling. First, we demonstrate doping/alloying energetics alongside a new thermodynamic stability model we developed for atomically-precise ligand protected nanoclusters that rationalize a series of experimental observations. [3-6] Second, we introduce a bond-centric model able to accurately describe alloy MNP stability of practically any MNP morphology and chemical composition. [7] Both examples make strides towards describing nanoparticle stability as a function of morphology and chemical composition, significantly accelerating nanomaterials discovery.

References

[1] M. G. Taylor, N. Austin, C. Gounaris, & G. Mpourmpakis, Catalyst Design Based on Morphology and Environment Dependent Adsorption on Metal Nanoparticles. **ACS Catalysis** 5, 6296–6301 (2015).

[2] J. Chung, I. Granja, M. G. Taylor et al. Molecular modifiers reveal a mechanism of pathological crystal growth inhibition. **Nature** 536, 446–450 (2016).

[3] M. G. Taylor, G. Mpourmpakis, Thermodynamic Stability of Ligand-Protected Metal Nanoclusters. **Nature Communications** 8, 15988 (2017).

[4] Q. Li et al., Reconstructing the Surface of Gold Nanoclusters by Cadmium Doping. **Journal of the American Chemical Society** 139, 17779-17782 (2017).

[5] Q. Li, K. J. Lambright, M. G. Taylor et al., Molecular "surgery" on a 23-gold-atom nanoparticle. **Science Advances** 21, 1-8 (2017).

[6] Q. Li, M. G. Taylor et al., Site-selective substitution of gold atoms in the $Au_{24}(SR)_{20}$ nanocluster by silver**. Journal of Colloid and Interface Science** 505, 1202-1207 (2017).

[7] Z. Yan, M. G. Taylor, A. Mascareno, G. Mpourmpakis, Size-, Shape-, and Composition-Dependent Model for Metal Nanoparticle Stability Prediction. **Nano Letters** 18, 2696-2704 (2018).

534205 Dissipative Particle Dynamics Simulations of Reactant Transport through Multicompartment Micelle Nanoreactor

Abstract:

Dissipative Particle Dynamics Simulation of Reactant Transport through Multicompartment Micelle Nanoreactor

SeungMin Lee, Connor Callaway, Nicholas Bond, Kayla Hendrickson, Aditya Kuntamukkula and Seung Soon Jang

Multicompartment micelles are a promising class of nanomaterials that is gaining interest due to potential applications in catalysis and drug delivery. The use of micelle nanoreactors as immobilized molecular catalysts is particularly enticing because it can combine the benefits of both high reactivity from homogenous catalysis and selectivity from heterogenous catalysis. Feasibility of the micelle as a catalyst is dependent on the stability of the micelle structure and the transport of the reactant molecules through the micelle to the different compartments.

Due to the complexity of the structure and slow process of experimentally studying reactant transport, we turn to dissipative particle dynamics (DPD) simulations to model the behavior of reactants in micelles. In our present work, we apply DPD techniques to produce micelles composed of linear terpolymers composed of decreasing hydrophobicity and vary the lengths of the block groups to obtain different micelle structures. One feature of the micelles to focus on will be the use of channel-like structures in the micelle which could facilitate reactant transport. Using the most stable and promising micelle structures for multi-step catalyzed reactions, we study the movement of reactants in the presence of these micelles to further support multi-step catalysis.

534640 Study of the Effect of a LiOH Layer over the Reactivity of Lithium Metal Anode

Abstract:

Electrochemical systems are highly dependent on interfacial chemistry where complex dynamic processes occur. In batteries, an interphase created between at the electrode/electrolyte interface of the lithium metal anode also called solid-electrolyte interphase (SEI), has received a lot of interest. The SEI is formed due to the decomposition of the electrolyte components upon electron transfer from the lithium metal surface, Depending on its composition and structure, the SEI can act as a passivation layer that can be desirable for the battery performance. One possible SEI component is LiOH, which is formed due to water decomposition. Even traces of water can form a thin layer on the Li metal surface. In this project, the impact of a pure LiOH thin film (~ 1nm thick) over the lithium metal anode is studied using first principles computational methods. Ab initio molecular dynamics (AIMD) and Bader charge analysis were performed to evaluate how this layer would alter the reactivity behavior of the lithium metal surface over the components of the electrolyte, and therefore the degree of passivation of the film. Also, partially covered LiOH surfaces are tested and discussed in this work

534829 The Influence of Pore Structure on Transport in Lyotropic Liquid Crystal Membranes

Abstract:

We study a lyotropic liquid crystal (LLC) system atomistically using molecular dynamics simulations. LLC's are a class of nanostructured materials that can be modified, assembled and crosslinked into stable membranes with uniform straight pores which can perform solute specific separations and offer the ability to control pore architecture at the atomic lengthscale. Pore diameters on the order of 1 nm make LLC membranes well-suited for aqueous separations such as desalination and biorefinement. A molecular model which is consistent with experimentally measured structural features and material properties can provide a link between monomer structure and macroscopic separation performance. We have used simulations with experimental data to create such a model. We have matched X-ray diffraction (XRD) patterns generated from equilibrated atomistic molecular dynamics simulation trajectories to experimental 2D wide angle X-ray scattering (WAXS) patterns in order to gain a detailed picture of the LLC membrane nanopores. We learned the density of monomers that pack around pore cylinders. We have evidence to support that aromatic monomer head groups Ï€-Ï€ stack on top of each other in a sandwiched, rather than parallel displaced, configuration. We show that the long alkane tails of each monomer pack into hexagonal arrays with short range order. We see that the pores are not idealistic cylinders, but rather consist of a gradient in composition transitioning from hydrophilic cores to hydrophobic tail regions. Using our detailed model of the membrane pores we are able to study the transport of molecules within them with confidence that the chemical environment closely models the real system. We discuss the transport rates and mechanisms of a series of neutral solutes with various hydrodynamic radii. With a clear understanding of mechanisms of transport in these complex self-assembled systems, one can choose monomers to achieve specific separation goals. We can use this information to draw correlations between pore structure and selective preferences. These studies will help guide monomer choice for separation-specific objectives.

535466 High-Throughput in silico Screening of Candidate Compounds for Deep Eutectic Solvents

Abstract:

Deep eutectic solvents (DES) are a class of generally low-cost, environmentally benign solvents that have diverse chemical applications. Similar in characteristics to ionic liquids (IL), they contain a mixture of cations and Lewis acid anions, which form the hydrogen bond acceptor (HBA) species, and Lewis or Bronsted acids, which form hydrogen bond donors (HBD). The H-bond links between the donors and the acceptors allow for melting-point depression in the eutectic mixture and hence lower melting points compared to ILs. Additionally, their easy synthesis from inexpensive starting chemicals along with low vapor pressures and flammability make them attractive systems. Our work concerns the development of novel DESs for use in supercapacitors as well as of the underlying molecular modeling techniques. Key target properties our research focuses on are wide electrochemical windows (ECWs), high ionic conductivities, and low melting points, which can be tuned at the molecular level. We are employing our automated high-throughput screening infrastructure to characterize large-scale candidate libraries. We utilize a diverse set of (multi-scale) molecular simulation techniques that range from first-principles electronic structure theory, to classical molecular dynamics, to cheminformatics, to custom-built machine learning prediction models. We will detail our mining efforts of the resulting data sets to extract rational design principles for next-generation DES lead candidates.

535461 Modeling Side Chain Conformations of Bottlebrush Polymers from iSAFT Density Functional Theory

Abstract:

A new density functional theory (DFT) formalism is designed to study the phase behavior of bottlebrush polymers in inhomogeneous systems. Bottlebrush polymers are macromolecules with polymeric side chains. They are of great research interest because of their potential applications such as rheology modifiers, polymeric photonics and drug-delivery vehicles. However, the structural details of bottlebrush polymers in solution are poorly understood. Particularly, conformational properties of the polymeric side-chains, including their flexibility and degree to which they are stretched, have not been elucidated. To provide new insight, we have developed a new DFT that is based on inhomogeneous statistical associating fluid theory (iSAFT). Assuming an infinitely long and rigid backbone, the theory is capable of predicting density distribution for the side chain segments. The effect of implicit solvent on bottlebrush polymers predicted by the theory agrees qualitatively with simulation results. Further the theory allows us to study how the side chain conformation varies with explicit solvent, different chain length, composition and grafting density. This theory has the advantages of lower computational expense compared with simulation and more detailed structural information compared with other theories, thus leading to a powerful tool to model bottlebrush polymer systems.

535634 Web Applications for Rapid Characterization of Nanoporous Materials

Abstract:

High-throughput molecular simulations and, more recently, machine learning models have revolutionized materials selection through rapid exploration of a large design space. Even so, setting up these calculations often requires significant expertise and time. Evaluating the properties of a single material is not necessarily straightforward, either. We have developed an online open source tool to run common adsorption characterization calculations of porous materials, such as metal-organic frameworks (MOFs), in a user-friendly package. By distributing the code and documentation as a web application, we can reduce the friction for users to set up and understand the software and relevant parameters. Easy characterization methods facilitate rapid prototyping of different MOF designs, guiding simulation and experimental efforts away from poorly performing materials. We will discuss the methodology of the underlying characterization methods, the software infrastructure for code provenance and distribution, and examples of how these tools have already been useful for our research group.

536222 Agglomerate Formation with Polydisperse Primary ParticlesÂ in the Transition Regime

Abstract:

During flame synthesis of nanoparticles, polydisperse primary particles (PPs) are formed by surface growth[1] and/or coalescence.[2] At high temperatures, PPs coagulate in the free molecular regime producing fractal-like agglomerates.[3] When these agglomerates exit the high-temperature region of flame reactors or enter high-pressure environments (e.g. in combustion engines), they coagulate in the transition regime as their mean free path becomes comparable to their size.[4] Agglomeration dynamics at these conditions depend on the spread of the PP size distribution.

Here, an Event-Driven (ED) model for agglomeration of polydisperse PPs is employed to investigate the evolution of agglomerate structure and size distribution from free molecular to transition regime. The broad agglomerate size distributions formed in the free molecular regime narrow by coagulation in the transition regime, attaining their quasi-Self-Preserving Size Distribution (SPSD) with gyration- and mobility-based geometric standard deviations of 1.62 ± 0.03 and 1.48 ± 0.03, respectively, regardless of PP polydispersity. Agglomeration results in a self-preserving scaling law between the mobility diameter, d_m, and the number of constituent PPs within an agglomerate, n_p. The ED-derived quasi-SPSD of the agglomerate d_m and its self-preserving scaling to n_p are in excellent agreement with mass-mobility measurements of soot and zirconia agglomerates and can be interfaced with method of moments or population balance models to assist the design of aerosol flame reactors and combustion engines.

References:

[1] Kelesidis, G. A.; Goudeli, E.; Pratsinis, S. E., Flame synthesis of functional nanostructured materials and devices: Surface growth and aggregation. *Proc Combust Inst* **2017,** *36*, 29-50.

[2] Eggersdorfer, M.L.; Kadau, D.; Herrmann, H.J.; Pratsinis, S.E., Multiparticle Sintering Dynamics: From Fractal-Like Aggregates to Compact Structures. *Langmuir* **2011**, *27*, 6358-6367.

[3] Goudeli, E.; Eggersdorfer, M.L.; Pratsinis, S.E., Coagulation of Agglomerates Consisting of Polydisperse Primary Particles. *Langmuir* **2016**, *32*, 9276-9285.

[4] Kelesidis, G. A.; Goudeli, E.; Pratsinis, S. E., Morphology and mobility diameter of carbonaceous aerosols during agglomeration and surface growth. *Carbon* **2017,** *121*, 527-535.

536316 Coarse-Grained Model for Simulating the Boiling Point of Asphaltenes

Abstract:

The reason behind computational chemistry is to use simulations to calculate the physical properties of substances in order to refine experimentation. For instance, a common problem in oil fields are Asphaltenes, a large hydrocarbon that has a tendency to clog the pipes and flowlines in oil wells. Determination of physical properties, such as the vapor-liquid equilibrium, would aid in also determining the necessary conditions for the aggregation of these compounds. An effective method using Wang-Landau approach combined with hybrid Monte Carlo simulations to calculate the boiling point and heat of enthalpy in the isothermal-isobaric ensemble. The method, HMC-WL simulation, was performed on a coarse-grained model of a simple asphaltene structure made of two pyrene groups linked by an alkyl chain of lengths from a chain of four methylene groups up to 16. Through the simulation a quasi-linear relationship was determined between the boiling point and the length of the chain.

536471 DFT Study on the Catalytic Activity of ALD-Grown Iron Oxide Nanoclusters for the Partial Oxidation of Methane to Methanol

Abstract:

The current route for converting natural gas into more useful chemicals is through complex, multi-step industrial processes that requires multiple reactors, separations, and high temperatures and pressures. Methane monooxygenase (MMO) is an enzyme found in biological systems that is capable of oxidizing methane to methanol using O_2 at mild conditions. Using density functional theory, we have studied the catalytic activity of ALD-grown iron oxide nanoclusters that mimic the structure of the active site in MMO for the oxidation of methane to methanol. Two mechanisms for the oxidation of methane are proposed using N_2O as the oxidant and a bridging oxygen (O_b) or a terminal oxygen (O_t) as the active site: (1) a concerted mechanism in which the C-H bond is cleaved heterolytically and (2) a rebound mechanism in which the oxygen atom abstracts a hydrogen atom from methane.

We have found that the radical rebound pathway is preferred over the concerted pathway by 40-50 kJ/mol. The desorption of methanol and the regeneration of the oxygen site exhibit high barriers for the direct conversion of methane to methanol. As demonstrated by a population analysis, the O_x (x= b or t) site behaves as an oxygen radical during the H-abstraction, and the $[Fe^{+}-O_x^{-}]$ site behaves as a Lewis-acid base pair during the concerted C-H cleavage. Molecular orbital decomposition analysis further demonstrates the electron transfer during the oxidation and reduction steps of the reaction. Understanding how these systems behave during the proposed reaction pathways provide new insights into how they can be tuned for alkane oxidation reactions.

536973 Conformational Mapping of Viral RNA Elements Using Atomistic Simulations

Abstract:

RNA molecules, similar to proteins, are highly flexible molecules that can undergo conformational changes in response to environmental fluctuations or due to binding of ligands. These conformational changes alter their function and can activate or terminate processes that are vital to lifecycles of cellular machinery. Viral RNA molecules are a good example of conformationally adapting molecules that have evolved to switch between many functions such as translation and replication. Specifically, the replication of human immunodeficiency virus Type-1 (HIV-1) results from a conformational change in the RNA transactivation response element (TAR RNA) elicited by the binding of the Tat protein. Since the discovery of TAR RNA, it has been explored as a potential target for inhibitors designed to block HIV-1 replication. However, the conformational dynamics of TAR-RNA with and without ligands/inhibitors has not been exhaustively sampled to date. In this work, we will present results from structure-based modeling and enhanced sampling methods to provide a detailed mapping of the conformational landscape of HIV-1 TAR RNA.

537040 A Molecular Dynamics Study on Interfacial Properties of $NaClO_4$/Carbonate Electrolyte Near Graphene-Based Electrode for Na-Ion Battery

Abstract:

Lithium ion batteries (LIBs) have been widely adopted as a promising power source for portable energy storage due to their high storage capacities and excellent rate performance over other rechargeable batteries. However, the cost for Li continues to be on the rise because of the increasing demand for LIBs. The abundance and low cost of sodium makes Sodium-ion batteries (NIBs) as a highly attractive alternatives for large-scale energy storage applications. Considerable efforts have been made to find suitable electrode and electrolytes materials for NIBs, However, the details of the electrode-electrolyte interface are not well-understood yet. In this work, molecular dynamics simulations have been employed to investigate the structural rearrangement of electrolyte comprised of ethylene carbonate(EC), dimethyl carbonate(DMC), and $NaClO_4$ salt near the pristine/N-doped graphene using Constant Potential Method (CPM). In the CPM simulations, the electrode charges are recalculated every time step and allowed to fluctuate in response to the local charge environment, which provides more realistic results compared to the simulations where the electrode atom charges are kept fixed. Our results show that the interfacial microstructure is highly dependent on the molecular geometry and polarity of each solvent and the electrode potential. Upon charging, the more polar EC is attracted while the less polar DMC is repelled from the interface. The results also show that ClO_4 anions are rapidly accumulated (repelled) near the positive (negative) electrode with increasing potential. On the other hand, Na+ in the interface has weak dependence on the applied potential excluding at very high negative potentials. Moreover, we found that the applied potential have a significant impact on composition of the electrolyte at the interface and solvation structure around Na cation. Quantitative understanding of the electrode-electrolyte interface learned from this study may bear important implications towards the early stages of solid electrolyte-interface (SEI) and intercalation/deintercation process of Na cation in the NIBs.

537129 Gating Mechanisms during Actin Filament Elongation By Formins

Abstract:

Formins are large multidomain proteins that play important role in the polymerization of actin filaments and can slow down polymerization by 5 to 95% depending on the particular type of formin. Our aim in this work is to identify factors that influence polymerization of actin by examining three types of formins with different actin polymerization rates, namely mDia1, Bni1 and Cdc12. We use multiscale computational modeling approach and enhanced free energy sampling simulation techniques to examine the interactions between each of these formins and an actin filament. The simulations reveal key dissimilarities in the structures and intermolecular interactions of the three formin-actin complexes. We identify two factors that influence the rate of formin mediated actin polymerization. Firstly, steric clashes at the barbed end of the actin filament block the addition of new actin subunits. Secondly, interactions between formins and actin filament affect the configuration of the barbed end by flattening the helical twist of the terminal actin subunits. Larger distortion in the terminal actin subunits makes the barbed end less favorable for subunit addition. We discover that the ability of each type of formin to slow down actin filament elongation correlates with both its degree of flattening of the barbed end and steric blocking.

537154 Comparison of Interatomic Potentials for Interfacial Studies of Ionic Liquid Systems

Abstract:

Molecular simulation is now a well-established too for understanding the structure, dynamics, and thermodynamics of ionic liquids at interfaces and in confinement. Some theoretical issues, however, are pervasive in the literature, largely owing to the selection and application of force fields. The most popular classical force field, the so-called CL&P force field of Canongia Lopes and Padua, has well-documented issues including slow transport properties and underprediction of hydrogen bonding behavior. Various corrections to the original force field have been explored and other similar force fields have been developed. The improvements are largely evaluated on the accuracy of their fundamental physiochemical properties, i.e. density, molecular diffusivity, enthalpy of vaporization, etc. Unclear, however, are the effects on high-level properties that drive the performance of ionic liquid applications. Additionally, the lack of a general protocol for handling fluid-solid interactions, controlling for any errors in the fluid-fluid interactions, provides further of uncertainty in the validity of these molecular models. To this end, we consider two systems that attempt to bridge low- and high-level properties: nanodroplets of ionic liquids on a graphene surface and a channel of ionic liquids interacting with charged graphene surfaces. In droplet simulations, we observe remarkably different wetting behavior across the different force fields and fluid-solid mixing rules. In channel simulations, we observe largely similar differential capacitances, despite notable differences in the structure of the ions at the interface and dynamics everywhere. These results imply that large inaccuracies in fundamental molecular properties may still be able to reasonably predict broader physical properties at larger scales.

537582 Mechanism Development for Catalyzed Ketene Production

Abstract:

A mechanism of pericyclic reactions involving metaphosphoric acid has been developed for the dehydration of acetic acid to form ketene using an organophosphate catalyst. Ketene is an important and highly reactive chemical precursor, including usage for the production of acetic anhydride. Dehydration of acetic acid with organophosphate catalysts has been observed and cited in industrial patents. The only elementary reaction mechanism of this sort that has been published was a study of catalysis by P_2O_3/P_2O_5 was published by Sebbar et al. (2016), addressing the reaction as a primarily radical reaction mechanism. This current work approaches the dehydration of acetic acid using triethylphosphate as the catalyst and considering only pericyclic reactions and demonstrates that the reaction proceeds through formation of the reactive intermediate metaphosphoric acid (HPO_3).

Stable intermediate species and transition-state structures included in the proposed mechanism were simulated using computational quantum chemistry software (Gaussian 09). Initial structure exploration and optimization were carried out using DFT methods at a B3LYP/6-31G(d,p) level of theory. Refined structures and their thermochemical properties were calculated with the composite method CBS-QB3. Hindered-rotation anharmonicity was considered for hydroxyl and methyl rotations in the structures using rotational energy profiles calculated at a B3LYP/6-31G++(d,p) level of theory. For electronically barrierless reactions, the transition state was calculated as the highest-free-energy structure in the reaction path, according to variational transition-state theory. Variational transition states were initially found using a B3LYP/6-31G++(d,p) level of theory and then refined using CBS-QB3. Rates of reaction were calculated using the Mesmer master equation code. Flow-reactor predictions using the resulting mechanisms and Chemkin Pro were compared against experimental flow-reactor yields as determined by molecular-beam mass-spectrometry. Potential pathways were evaluated by comparing reaction kinetics at the high-pressure limit against what was observed experimentally.

Eight possible pathways were considered. Only one pathway had rates fast enough to explain the experimental yields, the pathway involving metaphosphoric acid as a reactive intermediate. This mechanism begins with formation of phosphoric acid from triethylphosphate by successive eliminations of ethylene. Phosphoric acid then goes through a catalytic cycle in which it dehydrates to form metaphosphoric acid, associates with acetic acid to form acetylphosphate, and finally dissociates to yield ketene and a reformed phosphoric acid molecule. Predicted yields agreed with experimental yields within an order of magnitude without requiring any parameter fitting.

The authors gratefully acknowledge funding of this research by the Eastman Chemical Company and computational resources provided by the Extreme Science and Engineering Discovery Environment (XSEDE), which is supported by National Science Foundation grant number ACI-1548562.

538166 Foyer: A Framework for Defining Force Field Usage Semantics and Atom-Typing Molecular Systems

Abstract:

The availability of force fields for molecular simulation has reduced the effort researchers must devote to the task of determining the interactions between species, allowing them to instead focus on the motivating scientific questions. However, determining which parameters in a forcefield to use is still often a very tedious and error prone task, given that force fields may contain hundreds of unique parameters for a single element, differentiated by the chemical context for which they apply. While tools for performing atom-typing exist (i.e., determining which parameters describe an atom's interaction potential with its environment based on its chemical context), they are often tied to a specific force field or simulation software and typically employ a rigid hierarchy where rules that identify more specialized atom types must be called ahead of more general atom types. More importantly, many of these atom-typing tools are not directly tied to the files containing the definitions of the force field parameters, which may introduce ambiguities as to appropriate usage and the source of parameters. The strong dependence of force field parameters on the chemical context of the atoms demands a codified, logic-based expression tied directly to those parameters.

We present a tool - Foyer [1,2] - that eliminates ambiguity in force field parameter usage thereby enhancing reproducibility and simplifying the process of disseminating new force fields. Foyer provides a force field agnostic method for defining parameter usage that relies upon SMARTS [3] based annotations of chemical context to perform atom-typing and overrides statements to set rule precedence. In Foyer, rigid rule hierarchies are replaced by an iterative process which allows rules to be embedded in any order within the actual force field files alongside the parameter definitions, thus delivering annotations that are both human and machine readable. To demonstrate Foyer and how it can be used to simplify the process of disseminating a force field, providing automated testing, and ensuring that end users correctly use the derived parameters, we provide a case study for a novel force field for perfluoropolyethers [4]. Foyer has been developed as a Python library, designed to integrate with the Molecular Simulation Design Framework (MoSDeF) [5] currently under development at Vanderbilt.

1. Iacovella, C. R.; Sallai, J.; Klein, C.; Ma, T. Idea Paper : Development of a Software Framework for Formalizing Forcefield Atom-Typing for Molecular Simulation. In *4th Workshop on Sustainable Software for Science: Practice and Experiences (WSSSPE4)*; 2016.
2. "Foyer." [Online]. Available: https://github.com/mosdef-hub/foyer.
3. http://www.daylight.com/dayhtml/doc/theory/theory.smarts.html
4. Black, J. E.; Silva, G. M. C.; Klein, C.; Iacovella, C. R.; Morgado, P.; Martins, L. F. G.; Filipe, E. J. M.; McCabe, C. Perfluoropolyethers: Development of an All-Atom Force Field for Molecular Simulations and Validation with New Experimental Vapor Pressures and Liquid Densities. *J. Phys. Chem. B* **2017**, *121*(27), 6588–6600.
5. https://github.com/mosdef-hub

538250 MoSDeF: A Python-Based Molecular Simulation and Design Framework

Abstract:

The Molecular Simulation and Design Framework (MoSDeF) [1] is an extensible set of Python tools focused on designing chemical systems, atom typing, and screening over large sets of such systems in a programmatic fashion. MoSDeF has been developed as an effort to improve the current state of reproducibility within the molecular simulation community. Currently, MoSDeF is built around two key pieces of software, *mBuild* [2,3] for designing systems of interest, and *foyer* [4,5] for applying the appropriate force fields to a molecular system. *mBuild* supports a wide range of systems due to the robust particle connection method used and ability to declaratively express repetitious structures with small amounts of code. This leads to parameterization of many features of a system, for example, from terminal group chemistry of a polymer chain, basis particles within a crystal lattice, size and composition of a nanoparticle, and much more. *Foyer* allows for atom typing of molecular systems and does not rely on rigid, hard-coded, rule hierarchies for atom type application. By separating the engine that applies the force field from the force field file itself, the atom typing process can be debugged much easier, force fields can more easily be extended, evolved, and disseminated with publications. Additionally, in combination with the workflow management tool Signac-flow [6], MoSDeF allows for large-scale screening capabilities, while also maintaining reproducibility and human-readability.

References

[1] "MoSDeF" [Online]. Available: https://github.com/mosdef-hub.

[2] C. Klein, J. Sallai, T. J. Jones, C. R. Iacovella, C. McCabe, and P. T. Cummings, "A Hierarchical, Component Based Approach to Screening Properties of Soft Matter", Foundations of Molecular Modeling and Simulation, **2016**, pp. 79-92.

[3] "mBuild" [Online]. Available: https://github.com/mosdef-hub/mbuild.

[4] Iacovella, C. R.; Sallai, J.; Klein, C.; Ma, T. "In Idea Paper: Development of a Software Framework for Formalizing Forcefield Atom-Typing for Molecular Simulation", 4th Workshop on Sustainable Software for Science: Practice and Experiences (WSSSPE4), **2016**.

[5] "foyer" [Online]. Available: https://github.com/mosdef-hub/foyer.

[6] "Signac-flow" [Online]. Available: https://bitbucket.org/glotzer/signac-flow.

538451 Challenging Statistical Mechanics Approximations in Organic Crystal Thermodynamics

Abstract:

The physical and chemical properties of solid materials is largely influenced by the unique packing of the material, which becomes problematic when multiple solid forms, or polymorphs, exist. The particular packing of small organic molecules is of interest in fields such as as pharmaceutical formulation, organic electronics processing and explosives preservation. In recent decades, the focus has been on providing fast and efficient methods to predict favorable structures at any temperature of interest. When determining potential structures with computational methods the entropic and temperature effects cannot be neglected. Much of the work in crystal structure predictions are focused on predicting the static lattice energy landscape, which largely neglects the energy contributions due to crystal vibrations and thermal expansion.

The most accurate method to model organic material is through molecular dynamic simulations (MD), which account for all vibrational and thermal energy of the crystal lattice. With MD, our approach has developed a methods to compute the free energy difference between two crystal structures at a state point of interest, which can then be used with sample simulations at other state points to determine the free energy landscape at all temperature and pressures of interest. Alternatively, lattice dynamic (LD) approaches provide a less computationally demanding approach to determine the free energy of solid materials. We have also developed an approach to more efficiently to perform LD approaches, which now provides an easy approach to also perform anisotropic expansion.

These approaches have varying levels of accuracy and speed. The harmonic approximation (HA) uses only the lattice minimum harmonic vibrations to quickly compute the free energy. The drawback is that the HA does not account for thermal expansion and since all organic crystals expand with temperature most people use the quasi-harmonic approximation (QHA). Our new method for the QHA computes the gradient of thermal expansion of the crystal for either isotropic or anisotropic expansion to determine how the crystal geometry is altered as the temperature increases. Unfortunately, all three of these LD approaches fail to capture anharmonic vibrations, which effect the crystal energy at higher temperatures. To capture the full ensemble of crystal vibrations we compute the free energy differences between two crystals by moving each crystals along a reversible thermodynamic path to an ideal gas state at a single temperature point. Using sampled points at varying temperatures, we can compute the free energy differences between crystal polymorphs at all temperatures of interest.

538555 Molecular Modeling of Microstructure and Solubilization of Single and Multiple Micelles

Abstract:

Block copolymer micelles have been the interest of both academia and industry over decades for its extensive applications in micellar filtration, cosmetic products, and drug delivery. A few chemical potential models have been developed to predict sizes, aggregation number, solubilization ability, critical micelle concentration(CMC) etc. Most of the theories are phenomenological equation of state types of theories and thus cannot fully capture the physics at molecular level.

Interfacial Statistical Associating Fluid Theory(iSAFT) in this work is demonstrated to predict different influences of hydrophobe and hydrophile of block copolymer surfactants on the size and CMCs of a single micelle. Current theory can successfully predict micelles formed from triblock and diblock copolymers. These results agree with the experimental results of non-ionic micelles. We also studied the distribution of solute, solvent and surfactants within micelles using iSAFT. By taking advantage of a novel solution method, we also present results for semi-dilute self-assembled systems that contain multiple micelles and forms crystal-like structure.

538573 Proteins in Extreme Environments: From Understanding Life to Potential Applications

Abstract:

The traditional paradigm of biology often considers life over a very narrow range of physiological conditions. However, from the frigid waters of the Antarctic Ocean to high-altitude Andes, deep-sea hydrothermal vents within the Earthâ€™s crust or high-pressure depths of the sea, life can exist in all sorts of extreme environments. In fact, some organisms do not just survive but thrive in such environments. The existence of such extremophiles motivates the following questions: What are the fundamental features of life? What environments and evolutionary paths enable the survival of living organisms? Can we utilize our fundamental understanding of life in extreme environments for practical applications?

We use molecular simulations combined with advanced sampling methods and dimensionality reduction techniques to shed light onto how extreme conditions affect proteins - one of the major building blocks of life. Specifics areas of interest include ionic-liquid induced effects on protein folding mechanisms and kinetics, protein stability in aqueous inorganic solvents with direct applications in novel drug encapsulation methods, and the phase behavior of proteins as a response to protein-water interactions in the temperature-pressure space.

538813 Applications of Chemml Program Suite in Predicting Properties of Organic Materials: A Path to Data-Driven Discovery in Chemistry

Abstract:

Data mining and machine learning (ML) algorithms have received increased attention in chemical and materials research in recent years. To facilitate the broader dissemination of these techniques, we have developed an open-source program suite, called ChemML. Recent technical advances in ChemML allow us to uncover hidden structure-property relationships that govern the behavior of chemical and materials systems. These insights are a prerequisite for rational design and inverse engineering capability as outlined in the White House Materials Genome Initiative. This presentation highlights capabilities of ChemML for accurately predicting properties of organic molecules in different applications. For a benchmark data set of electronic properties of more than 2 million organic semiconductors, we demonstrate substructure-based fingerprints allowed for the construction of ML models with state-of-the-art performance. We also employ ChemML to predict refractive index of polymers through different deep learning architectures. These models merge different feature representation of polymers and using design methodologies achieve predictions that are competitive with significantly more complex machine learning techniques. All these models can further be used in virtual high-throughput screening for accelerating materials discovery and as a guide for rational design.

521428 Benchmarks for Adsorption on Transition Metal Oxide Surfaces: A Comparison of DFT to Experimental Data for NH_3 on MnO(100)

Abstract:

Density functional theory (DFT) is commonly used to investigate adsorption and reaction processes on surfaces. While it is widely used, the applicability of DFT to highly correlated transition metal oxide systems with unpaired d-electrons is problematic, especially when using plane wave basis sets. A primary issue is the lack of quality experimental data for adsorbate and reaction systems on well-defined (single crystal) transition metal oxide surfaces to provide benchmarks for the applicability of current DFT methods and the future development of new methods. We have established experimental benchmarks using temperature programmed desorption (TPD) for the adsorption energy of ammonia molecules (NH_3) on MnO(100) for comparison to DFT simulations on terraces, using Perdew-Burke-Ernzerhof (PBE) functionals with DFT+U, in conjunction with a number of Van der Waals force correction packages.

522681 Combining Molecular Simulation, Liquid State Theory, and Gibbs Ensemble Techniques to Study the Structure, Thermodynamics, and Phase Behavior of Polymer-Solvent Mixtures

Abstract:

Solvent processing techniques (*e.g.* solvent vapor annealing, flash nanoprecipitation, reverse emulsion assembly, various thin-film casting and coating processes) play a vital role in the fabrication of functional polymer materials. In each of these examples, the presence of solvent alters polymer thermodynamics by screening polymer-polymer interactions and tunes polymer dynamics by increasing chain mobility, phenomena which can be exploited to trigger some structural or morphological change in a polymer material. However, in order to design effective solution processing methods, there is a need for a fundamental understanding of how varying *solvent content* (dilute, semidilute, concentrated) and *solvent quality* (poor, theta, good) affect the structure and thermodynamics of polymer solutions. In this poster, we describe our recent developments of computational techniques to study the solvent processing of polymer systems. First, we present a hybrid Monte Carlo/molecular dynamics (MC/MD) Gibbs ensemble approach to study vapor-liquid equilibria in polymer-solvent systems. We apply this technique to the solvent vapor annealing of polymer blends to uncover the impacts of changing *solvent content* on blend structure and thermodynamics. Second, we detail a combined simulation and theory approach using coarse-grained MD simulation and Polymer Reference Interaction Site Model (PRISM) theory to study polymer chain conformations and polymer-solvent thermodynamics over a range of *solvent qualities*. Thus, using our computational techniques, we can provide a useful mapping of the structure and thermodynamics of polymers in varying solution conditions to inform the design of solvent processing methods.

525641 Molecular Dynamics Simulation of Hydration and Swelling of Mixed Layer Clays

Abstract:

Swelling of clay minerals is important to a broad class problems in science and engineering. While the problem has been extensively studied experimentally, past molecular modeling of the phenomenon was focused on pure clays of one type or another. In practice, however, there is a diverse class of mixed-layer clays (MLCs) in sedimentary rock with inter-mixed stacking sequence of two or more types of distinct layers within a single crystal. In fact, more than 60 percent of sedimentary rocks in the U.S. contain various types of MLCs. We present the results, to our knowledge, of the first molecular dynamics simulation of hydration and swelling of illite-montmorillonite (I-MMT) MLCs, the most common type of such clays. The swelling is studied as a function of water concentration with four combinations of inter-layer cations, Na+ and K.+ The hydration energy, radial distribution functions, and the density profiles in the inter-layer region are computed. For regular Na-MMT with layer charge concentrated in the octahedral sheet, weak cation-surface interaction results in fully hydrated ions an significant swelling. In the asymmetric inter-layer of the MLC, however, the illite sheet with
stronger interaction of surface and ions causes adsorption of the cations deep in the ditrigonal cavities of the siloxane surface. Given that the hydration enthalpy of K+ is smaller than that of Na,+ its hydration shell is quite unstable compared with that of Na.+ Therefore, swelling is inhibited as the ratio K+/Na+ increases. The results demonstrate the significant differences between the hydration and swelling properties of pure clays and the mixed ones, which have important implications in practice.

534098 Colloidal Crystal Structure Analysis Using Symmetry Groups and Stochastic Optimization

Abstract:

We present PACCS, a Python package for the analysis of colloidal crystal structures and prediction of the stability of such structures under various conditions. PACCS uses symmetry groups, specifically, the seventeen unique wallpaper groups in two-dimensional space, to generate various periodic arrangements of particles. These configurations are then used as inputs to a stochastic optimization algorithm such as differential evolution or basin-hopping to generate a library of distinct candidates from which a ground state (minimum energy) structure can be found. Free energy calculations can also be performed on these lattices to evaluate stability of phases at finite temperature. We discuss how the symmetries of a selected wallpaper group and the requirements of a given crystal stoichiometry can be cast as a constraint satisfaction problem for solution and generation of these initial inputs. We consider how this approach functions for a simple colloidal system in two dimensions, and also show how various complexities can be handled: size disparate systems, molecular systems, and three-dimensional systems. We demonstrate that this algorithm is amenable to parallelization and compare its performance to that of stochastic optimization alone, showings its usefulness for enumerating structures for multicomponent mixtures in a computationally efficient manner.

525278 Feasst: Free Energy and Advanced Sampling Simulation Toolkit

Abstract:

The Free Energy and Advanced Sampling Simulation Toolkit (FEASST) is a free, open-source, modular program to conduct molecular and particle-based simulations with Metropolis, Wang-Landau and Transition-Matrix Monte Carlo methods [1-2]. FEASST is implemented in C++ and may be imported as a module within Python 2 or 3. FEASST is capable of simulating systems in the (semi-)grand canonical, isothermal-isobaric and canonical ensembles. The following potential models are also available: Kern-Frenkel patchy particles, screened and long-range electrostatics with Ewald, Lennard-Jones, square well and tabular potentials. Many more features are planned for release including expanded ensembles in temperature and alchemical transformations, configurational bias, geometric cluster algorithm, aggregation volume bias, confinement and Mayer-sampling Monte Carlo.

[1] H.W. Hatch, et al., J. Res. Natl. Inst. Stan., 123, 123004 (2018). https://doi.org/10.6028/jres.123.004
[2] https://pages.nist.gov/feasst/

538844 Modeling the Transformation of Ethene over MFI Using a Hybrid QM/MM Strategy

Abstract:

The conversion of ethene into larger intermediates is an important step in the dehydroaromatization scheme involving the conversion of ethene into aromatics over MFI Zeolites. However, to date, a detailed investigation of the mechanism of ethene conversion including entropic effects has been lacking in the literature. In this study, we employ a hybrid Quantum Mechanics/ Molecular Mechanics Approach (QM/MM) which uses a high quality DFT functional and a triple zeta basis set for the QM region, where the MM region has been parameterized to accurately capture the effects of molecular adsorption within the MFI lattice for both polar and nonpolar adsorbates. Alkoxide-mediated C-C coupling mechanisms are investigated on free energy landscapes, and a detailed comparison with experimental data is to be made.

539123 Prediction of Phase Behavior of Mixed Solvent Electrolyte Systems Using SAFT-VRE Morse EoS

Abstract:

In this study, the variable range version of electrolyte Statistical Associating Fluid Theory (SAFT-VRE) equation of state (EOS) has been considered to represent the phase behavior of mixed solvent electrolyte solutions. In this regard, the SAFT-VR Morse is considered for short-range dispersive interaction whereas the Debye-Huckle term is utilized for long-range electrostatic interactions between charged particles. The ion-specific parameters are obtained by the fitting to experimental mean ionic activity coefficient and solution density data of several 1:1, 1:2 and 2:1 types binary electrolytes. The thermodynamic properties and phase behavior of mixed solvent electrolyte solutions are predicted using the obtained ion-specific parameters in binary electrolyte solutions (salt-water solutions). It is well documented that the knowledge of dielectric constant of water-cosolvent mixture is essential to predict the thermodynamic properties of aforementioned systems. In this regard, a simple model is used to calculate the mixed solvent dielectric constant. The results show that, predictive capability of SAFT EoSâ€™s could be improved using the suitable variable range dispersion term such as SAFT-VR Morse model.

521331 Characterization of Heat Absorption and Decomposition Products for Suppressant Agent/Combustible Dust Mixtures Via TGA/DSC/MS Analysis

Abstract:

Combustible dusts continue to present a critical threat towards personnel and process equipment in a wide variety of bulk solids processing industries. Addition of non-combustible inert material to combustible dust mixtures, either through pre-mixing or injection as the deflagration begins to develop, is common practice for preventative inhibition or explosion protection (mitigation) via active suppression, respectively. Metal dusts specifically present an extremely reactive explosion risk due to characteristically augmented heats of ignition, burning temperature, flame speed, explosibility parameters (K_{St} and P_{max}), and ignition sensitivity. Establishment of proper prevention and protection designs for metal fuels has posed challenging. The aim of this paper is to propose a method for the characterization of the inhibition efficiency of six suppressant agents (sodium bicarbonate [SBC], potassium bicarbonate [PK], monoammonium phosphate [MAP], diammonium phosphate [DAP], potassium allophanate-based [Agent â€˜Zâ€™], and sodium chloride-based [Met-L-X^{TM}]) when mixed with fuels of both organic (cornstarch) and metallic (zinc) nature. For each 1:1 wt% fuel/agent mixture, thermal stability and heat absorption efficiency have been evaluated through thermogravimetric analysis (TGA) and differential scanning calorimetry (DSC).

527920 Measurement of the Liquid Thermal Conductivity of HFO-1336mzz(Z)(*cis*-1,1,1,4,4,4-hexafluoro-2-butene) By Transient Hot-Wire Method

Abstract:

With the increasingly prominent environmental problems, the HCFCs refrigerants are now being accelerated eliminated, and all of them have been banned and gradually replaced by the refrigerants with zero ozone depletion potential (ODP). As one of the hot topics in the field of refrigeration and air conditioning, the development of zero ODP and low Global Warming Potential (GWP) environmental refrigerants has become an urgent problem in the refrigeration and HVAC industry, and has been widely concerned by scholars both at home and abroad. Hydrofluoroolefins (HFOs) are considered as the new generation of refrigerants and working fluids because of their zero ODP and low GWP. HFO-1336mzz(Z) (*cis*-1,1,1,4,4,4-hexafluoro-2-butene, *cis*-$CF_3CH{=}CHCF_3$, CAS NO.692-49-9) has been discussed recently because of its nonflammability and low GWP value. In addition, HFO-1336mzz(Z) has excellent thermodynamic and chemical stability which is considered as a potential working fluid for various applications, including high temperature heat pumps, Organic Rankine Cycles and commercial chillers. Furthermore, HFO-1336mzz(Z) can also be used as foaming agent, extinguishing agent and solvent. In the existing literature, the thermal conductivity data of HFO-1336mzz(Z) are very few. Accordingly, in this work, the thermal conductivity of HFO-1336mzz(Z) for the compressed liquid phase were measured using the transient hot-wire thermal conductivity measurement equipment and basing on the Wheatstone bridge over the temperature range of (270 to 440)K with pressures up to 10 MPa. The transient hot-wire device was used to obtain high quality data for the thermal conductivity to help improve the design and operating costs of equipment item such as heat exchangers and provide reference for the other engineering application.

528777 Thermodynamic Properties and Molecular Interactions of Azeotropic Mixtures Using Molecular Simulation and Modeling

Abstract:

The occurrence of azeotropes poses great challenges for separation processes based on vapor-liquid equilibria (VLE), such as distillation, where the liquid and vapor phases have identical compositions. Despite the abundance of experimental data of many azeotropic systems, understanding into the molecular mechanism is lacking which is impeding the design of more efficient separation processes. In this study, we use the united atom transferable potentials for phase equilibria (TraPPE-UA) force-field to model, at the molecular level, different azeotropic systems with both positive and negative deviations (from the ideal solution). VLE of these mixtures are explored using Gibbs ensemble Monte Carlo (GEMC) simulation. Our VLE results agree well with experimental data. Partial thermodynamic properties of different components in the mixtures are calculated to reveal the molecular origin for the strong departure from ideal solution behaviors. Liquid structure of the mixture is characterized for the molecular details responsible for the thermodynamic changes. Finally, we attempt at the quantitative structure-activity (QSAR) modeling for the prediction of the existence of azeotropes from the molecular constituents based on data collected from both molecular simulation and experimental measurements.

530723 Refining the Nonrandom Two-Liquid Segment Activity Coefficient Model By Applying the Association Theory

Abstract:

Much interest has recently been devoted to accurately correlating and predicting drug solubility (Tanveer, S., et al., *Chem. Eng. Prog.* 2014, *110* (9), 37-47). Solubility models like the Non-Random Two Liquid Segment Activity Coefficient (NRTL-SAC) model (Chen, C.-C., and Song, Y., *Ind. & Eng. Chem. Res.* 2004, *43* (26), 8354-8362) are now routinely used in the pharmaceutical industry. However, this model does not explicitly account for specific intermolecular interactions such as hydrogen bonding. In this study, we refine the NRTL-SAC model by including an association term of activity coefficient, which is based on Wertheimâ€™s association theory. Additionally, we reduce the number of conceptual segments to two to represent polar and nonpolar surface characteristics of molecules. We also define two types of association sites to represent the hydrogen bond donor and hydrogen bond acceptor, respectively: molecules have both types of sites self-associate; molecules containing one type of sites can only cross-associate with molecules containing the other type of sites. The refined model generally performs better than the original model for various phase equilibrium calculations, especially for systems containing hydrogen bonds. The correlation and prediction of drug solubility are also improved by the refined NRTL-SAC model.

531558 Thermodynamic Description of Shear-Induced Phase Transition in Jammed Soft Particle Glasses

Abstract:

Recently we computationally showed that jammed suspensions of soft particles can undergo a phase transition under steady (Khabaz *et al. Phys. Rev. Fluids.* **2,** 093301 (2017)) and oscillatory (Khabaz *et al. Phys. Rev. Fluids.* **3**, 033301 (2018)) shear flows and form variety of ordered microstructures. In particular using the three-dimensional particle-dynamics simulation, we demonstrated that polydisperse suspensions formed a layered microstructure parallel to the flow-vorticity plane at high shear rates. The formation of the layered microstructure significantly decreases the internal energy and entropy of the system. Although the thermal forces are negligible in our model for jammed suspensions of soft particles due to the significance of the elastic forces between particles at contact, we can define an effective temperature using the excess entropy and elastic energy of the system. Rosenfeld theory, which connects the excess entropy to the microstructure of a system, is employed to quantify the excess entropy of the glassy and layered suspensions under different shear rates and volume fractions. Our results at a given shear rate show that the temperature and Helmholtz free energy of the layered microstructure is lower than that of the glassy suspensions. Both temperature and Helmholtz free energy follow a similar trend as the flow curve of these materials, i.e. an increase in the shear rate in each state increases the temperature and elastic energy of the suspension. In addition, the defined temperature exhibits an approximately linear relationship with the shear stress and elastic energy, which indicates that in a jammed state of these soft suspensions the shear stress can play the temperature role.

536728 CFD Simulations for Gas Solubility Measurements with Gas-Liquid Segmented Flows

Abstract:

Gas absorption in liquid solutions involves a multiphase mass transfer mechanism. The resistance present at phase boundaries controls the mass transfer rates and hampers the measurement of thermodynamic properties such as gas solubility and reaction rates. Gas-liquid systems in microfluidic scale overcome the limitation in the mass transfer rate by increasing the specific interfacial contact area across phases and reducing the mass transfer resistance [1]. Recirculation flows present in liquid slugs further improve mass transfer [2]. Mass transfer rate is faster in microfluidic scale which significantly reduces the timescale of measurement [3]. We study the solubility of CO_2 gas in water and in aqueous NaCl solution (at 298K) using Computational Fluid Dynamics simulations. CFD simulations allow accurate examination of pressure profile, velocity, and concentration fields in the system. We design a unit cell geometry in CFD which depicts a single unit of a bubble and slug train. Periodic conditions across the unit cell geometry are used which makes unit cell design a good approximation and significantly reduces computation time. The level set method coupled with momentum balance and continuity equation are used in CFD to capture fluid flow profiles. To address mass transfer in CFD, we use the convection-diffusion transport equation to capture the concentration profile. CFD simulation results show pressure profile, velocity profile, and concentration field which can be used to calculate the solubility. We show that CFD simulations coupled with gas dissolution experiments in gas-liquid segmented flows can identify the solubility of CO_2 in different solutions and reduce the experimental effort.

References

[1] Günther, A. and Jensen, K.F. "Multiphase microfluidics: from flow characteristics to chemical and materials synthesis." *Lab on a Chip*, 6(12), 1487-1503, 2006.

[2] Taylor, G.I. "Deposition of a viscous fluid on the wall of a tube." *Journal of Fluid Mechanics*, 10(2), 161-165, 1961.

[3] Yang, L. Nieves-Remacha, M.J. and Jensen, K.F. "Simulations and analysis of multiphase transport and reaction in segmented flow microreactors." *Chemical Engineering Science*, 169, 106-116, 2017.

537968 Thermodynamic Modeling of Saturn Particles and Phase Behavior in Patchy Colloid Mixtures

Abstract:

Research on patchy colloids has attracted ongoing interest and the fabrication of patchy colloids with different number, size, shape, and relative location of attractive surface patches makes patchy colloids a good candidate as building blocks to form much more complex advanced materials. Particles with Saturn ring like bonding sites have been fabricated in experiment and they can be used to stabilize Pickering membranes[1] . In this work we extend a cluster size distribution theory[2,3] to study the thermodynamics and phase behavior of binary patchy colloid mixtures. The binary mixtures contain solute colloids with a single Saturn ring like bonding site (Saturn solute) which can form multiple bonds with solvent particles. The solvent particles contain one or two association sites which are restricted to bond only once. The interactions between particles consist of hard-sphere repulsion and short-ranged directional attraction. We explain how accurate multiple-body correlation information is obtained from a cluster size distribution theory for solvent particles in the bonding region of solute. Then we show that it is feasible to extend the cluster size distribution theory to study a binary mixture of patchy colloids in which the solute can have a single geometrical association site that can bond multiple times or multiple single-bonding sites with different arrangement on the surface. By studying the binary mixture containing Saturn solute with one or two patch solvents, we show that the theory results match well with Monte Carlo simulation for the bonding state at different solute fractions, bonding energies and reduced densities. We also study how phase behavior is affected by the size of the solute Saturn ring, the solute fraction and the ratio of solute-solvent bonding energy to solvent-solvent bonding energy.

Reference:

1.Krejca, Matthias M., Cornell Wüstner, and Werner A. Goedel. 2017. "Pickering Membranes Stabilized by Saturn Particles." *Langmuir* 33(41): 10772–81.

2.Bansal, Artee, Walter G. Chapman, and D. Asthagiri. 2017. "Quasichemical Theory and the Description of Associating Fluids Relative to a Reference: Multiple Bonding of a Single Site Solute." *Journal of Chemical Physics* 147(12).

3.Bansal, Artee, Arjun Valiya Parambathu, and D Asthagiri. 2017. "Thermodynamics of Mixtures of Patchy and Spherical Colloids of Different Sizes : A Multi-Body Association Theory with Complete." : 1–30.

532669 PID Control Strategy for Thermostating and Barostating Molecular Dynamics Simulation

Abstract:

In molecular dynamics (MD) simulations, a number of temperature control methods (thermostating) have been employed for decades [1-4]. A well-known conventional method is the Nosé-Hoover thermostat [4], which acts like an integral control method. Within process systems engineering, the proportional-integral-derivative control (PID) is well established for decentralized single control loops. In this research, the authors have developed a PID framework (PIDStat) for MD thermostating. The PID formalism allows flexibility in MD thermostating. In NVT ensembles it is observed that through the suggested approach, the equilibrium time becomes shorter, while overshoots and fluctuations are less pronounced than in the Nosé-Hoover thermostat. PIDStat has been extended to pressure control (barostating). The authors are currently interfacing PIDStat to the well-known open-source MD code, LAMMPS for simulation of complex systems. The expectation is that thermostating and barostating can be accomplished within a shorter period of time, thus allowing for tackling larger problems. The advantages of PIDStat continues to be explored.

535172 Multiscale Modeling of Actin Filaments

Abstract:

Actin is an important protein in the cellular cytoskeleton, which polymerizes into polar filaments that form dynamic actin networks including filopodia, in which parallel actin filaments are bundled together. Each polar filament grows at its barbed end and shrinks at its pointed end under physiological conditions. The filament also ages as the nucleotide ATP bound to an actin subunit in the filament hydrolyzes and releases inorganic phosphate, modulating its mechanical properties and binding affinity towards several important actin binding proteins. Enabled/vasodilator-stimulated phosphoprotein (Ena/VASP) is an actin binding protein that assists filopodia formation by continuously associating with the growing barbed ends of predominantly short actin filaments in the bundles. The structure of a wild-type Ena/VASP loosely resembles a four-arm star polymer, with each arm consisting of several important domains responsible for interactions with actin.

In vitro microscopy experiments of purified actin filaments with Ena/VASP mutants of varying functionalities exhibit a rich phenomenology. A systematic multiscale modeling approach is used to provide insights into the experimental observations at the levels of individual arms and individual subdomains. A model based on kinetics of individual arms of Ena/VASP is constructed to understand the experimentally observed binding-unbinding rates of Ena/VASP mutants on actin filaments. Additionally, an Ultra-Coarse-Grained (UCG) model of actin is constructed, which provides a computationally efficient way to simulate long actin filaments undergoing ATP hydrolysis and growth. A systematic procedure is used to construct heterogeneous elastic networks for important domains of Ena/VASP and the actin filament including the implicit representation of subunit specific states of the bound nucleotide, using information from underlying atomistic simulations of short actin filaments interacting with Ena/VASP. Transitions between the three states of the nucleotide bound to each actin subunit are simulated in the UCG model via stochastic transitions coupled to the positions of the coarse-grained sites, and hence to the conformations of the underlying reference atomistic system. The resulting model takes into account detailed interactions at the subdomain level and is used to further study the growth of actin filament bundles in presence of Ena/VASP.

526580 Fluid Behavior and Interfacial Structure of Heterogeneous GO Interlayer Pores

Abstract:

Fluid Behavior and Interfacial Structure of

Heterogeneous GO Interlayer Pores

*Tongfei Yu, Shuyan Liu, Xiaoning Yang**

State Key Laboratory of Materials-Oriented Chemical Engineering,

Nanjing Tech University, Nanjing 210009, China

ABSTRACT: Water permeation through Graphene oxide (GO)-based laminate membrane has attracted increasing attention due to its great potential in nanofluidics and separation applications. However, the impact of complicated heterogeneous features of GOs on the water transport still remains largely unresolved and commonly neglected. In particular, recent various new-typed GO-based composite membranes further highlighted the importance of the hetero-structures. A comprehensive and in-depth understanding of the roles various structural features on water flow across the GO interlayer is significant and imperative for optimal design and application of GO-based membrane.

In this work, comprehensive MD simulations were performed to study the water transport and confined structure through a series of heterogeneous 2-D GO-based channels. Meanwhile, we theoretically developed a flow equation for heterogeneous 2-D channels assembled with asymmetrical GO sheets. This equation could reveal the leading role of the high-friction side of heterogeneous channels in controlling flow enhancement. By combining the MD simulation and theoretical analysis, it is found that the presence of one oxidized sheet could significantly prohibit the fast flow even though another sheet is frictionless pristine graphene, which is consistent with the theoretical prediction results. The distinct vertical drag effect stemmed from the interfacial affinity of GO surface has been revealed as the underlying mechanism, which is characterized by the asymmetric velocity profile and unique interfacial hydrogen bond behavior, as well as surface affinity. Our results provide new insight into the flow behavior between heterogeneous GO interlayer galleries and computationally reveal the unusual mechanism of the completely different water transport, as compared with that in traditional homogeneous pores. These unique understandings are not only valuable for guiding and improving the rational design of GO-based membranes, but also provide new physical pictures for heterogeneous capillary channel in extensive 2-D nanofluidics.

ACKNOWLEDGMENTS

This work was supported by the National Natural Science Foundation of China under Grants 21676136

*: Corresponding author, Yangxia@njtech.edu.cn

529636 Molecular Simulation and Experimental Study of Oxalic Acid Adsorption on Water-Feldspar Interface

Abstract:

Molecular simulation and experimental study of oxalic acid adsorption on water-feldspar interface

Xue Xiaopeng **(speaker)**, Li Ping*, Yu Jianguo

State Key Laboratory of Chemical Engineering, College of Chemical Engineering, East China University of Science and Technology, Shanghai 200237, China

*Correspondence author email: imusik@163.com, liping_2007@ecust.edu.cn

Abstract

Adsorption of organic molecules on mineral-water interface plays a significant role in many fields of application, such as minerals flotation with organic agents, bioleaching minerals by organic metabolites of microorganisms, interaction between organic acids and minerals for nutrient component acquisition in soils. Generally, there are main two kinds of mechanism for organic molecules adsorption on mineral surface, namely as inner-sphere adsorption and outer-sphere adsorption. For inner-sphere adsorption, there exists the direct chemical bond adsorption between organic ligands and cations of mineral surface. For outer-sphere adsorption, organic molecules will interact with hydroxyl groups on mineral surface through hydrogen bonding or electrostatic interaction.

During bioleaching feldspar to extract potassium element as chemical fertiliser, it is found that oxalic acid adsorption on water-feldspar interface is important for feldspar decomposition. Moreover, feldspar ($KAlSi_3O_8$) is also most common mineral and accounts for 98% in earth¡¯s crust, and oxalic acid is one of the most common microbial metabolites. In this work, both classical molecular dynamic simulation and DFT calculation are adopted to investigate oxalic acid adsorption on water-feldspar interface, and the predicted results are validated with the detected ATR-FTIR spectra.

Fig.1 Adsorption of $H_2C_2O_4$, $HC_2O_4^-$ and $C_2O_4^{2-}$ on water-feldspar interface

Three forms of oxalic acid interacted with feldspar are built, as shown in Fig. 1, at adsorption equilibrium states obtained through 1 ns molecular dynamic simulation. Adsorption energies for three system are calculated as 0.1 ev, 0.7ev and 1.3ev, respectively. According to the simulation results, it is found that $H_2C_2O_4$ is adsorbed by outer-sphere mechanism, while $HC_2O_4^-$ and $C_2O_4^{2-}$ are absorbed by inner-sphere mechanism.

As known, there are two sites, Al site and Si site, on feldspar surface for competitive adsorption of organic ligands, which site would be prior to be occupied by $HC_2O_4^-$ and $C_2O_4^{2-}$? Here, we shall adopt both DFT simulation and ATR-FTIR detection to recognize the characteristic bond frequencies, to validate oxalic acid adsorption on Al site, not on Si site.

Fig.2 ATR-FTIR spectra and DFT predictions of frequency.

The detected ATR-FTIR spectra for $H_2C_2O_4$, $HC_2O_4^-$, $C_2O_4^2$, feldspar, feldspar-$H_2C_2O_4$, feldspar-$HC_2O_4^-$, feldspar-$C_2O_4^2$, are shown in **Fig.2**, respectively, it is found two new peaks after adsorption of HC_2O_4 and $C_2O_4^{2-}$, one at 1617cm-^{1} and another at 1317cm^{-1}. Then, we make molecular simulation with HF/3-21g* and DFT basis: B3LYP/6-31g* to predict frequencies of Al-oxa and Si-oxa. It can be seen Al-Oxa frequencies are more close to the detected frequencies. Therefore, $HC_2O_4^-$ and $C_2O_4^{2-}$ are absorbed through interaction between O atom of $HC_2O_4^-$ ($C_2O_4^{2-}$) and Al atom rather than Si atom, which is inner-sphere adsorption mechanism. $H_2C_2O_4$ is adsorbed through electrostatic force between O atom of feldspar and H atom of $H_2C_2O_4$, which is outer-sphere adsorption mechanism.

531428 Molecular Dynamics Analysis of Membrane Proteins As Biosurfactants into Triglycerides – Water Mixtures

Abstract:

MOLECULAR DYNAMICS ANALYSIS OF MEMBRANE PROTEINS AS BIOSURFACTANTS INTO TRIGLYCERIDES – WATER MIXTURES

Cardona Jaramillo, J. E. C.[1]; Alvarez O.A.[1], Achenie, L.E.K. [2]; Gonzalez Barrios, A.F.[1]

1. Grupo de Diseño de Productos y Procesos (GDPP), Departamento de Ingeniería Química, Universidad de los Andes, Bogotá, Colombia {je. Cardona10, oalvarez, andgonza} @uniandes.edu.co
2. Department of Chemical Engineering. Virginia Polytechnic Institute and State University. achenie@vt.edu

Cosmetics are mass consumption products that play an important role in our daily lives. Therefore, the cosmetic industry is a rapidly growing economic sector, whose main research interest is staying ahead in a highly competitive market through product improvements [1]. As many finished products are made from colloidal mixtures, identifying surfactants that modify the interfacial tension between immiscible liquids to form emulsions is a very attractive research topic. The surfactant benefit and societal impact are increased if the surfactant is derived from a renewable source. Biosurfactants are a structurally quite diverse group of compounds synthesized by microorganisms, that are characterized by large structures which confer special properties such as surface activity [2]. Among these natural compounds, the membrane protein A from *Escherichia coli* (OmpA) has been recognized as a potential biosurfactant because it has a high homology with the emulsifier protein of the commercial biosurfactant Alasan (i.e. AlnA) [3], [4]. Considering that membrane proteins have an amphiphilic nature that allows them to interact both with the middle of the periplasm as well as the hydrophobic membrane, OmpA, OmpN and other membrane proteins may be potential biosurfactants [3], [4]. However, there are a poorly knowledge about the interaction patterns between biosurfactants and hydrophobic molecules.

A mixture of water and oil forms a model system that can lead to a better understanding of the biosurfactant behavior in a cosmetic product. Vegetable oils are commonly included as the main ingredients in cosmetic products, and they are composed mainly of triglycerides that are esters formed by glycerol and three fatty acids [5]. Those type of oils are complex mixtures of triglycerides with different unsaturation levels and chain lengths. The use of the oils requires a real understanding of the composition of the model system because it directly affects skin sensation during use. Also, depending on its composition, vegetable oils could be susceptible to degradation because the level of unsaturation in the fatty acid chain promotes oxidation reactions [5].

A computational strategy, specifically an all atom molecular dynamics (MD) for establishing the effect of two selected biosurfactants (i.e. OmpA and OmpN) within the interface. To model the vegetable oil, a set of different triglycerides was analyzed. In greater detail, a cross-factorial array experimental design was used

to analyze the effect of three different factors, namely the type of biosurfactant molecule, the presence of double bonds in the triglyceride molecule, and the triglyceride molar concentration. As explained above, OmpA and OmpN served as biosurfactants, while T54:3, T54:0, T48:3 and T48:0 were the triglycerides (where the number after "T" denotes the number of carbon atoms within the triglyceride molecule, and the second number represents the number of double bonds). Simulation boxes included one biosurfactant molecule, 6 or 12 triglyceride molecules and 3231 water molecules. MD were performed using LAMMPS (Nov. 17, 2016 version)[6]. Each system was energy minimized, and then equilibrated in a microcanonical ensemble. The production step was carried out with an a NVT ensemble for 1ns. All trajectories were analyzed using the Ovito [7] and VMD [8] packages through the estimation of the system's surface accessible to solvent area (SASA), second virial coefficient (β) and radial distribution function g(r).

Our results suggest differences in the biosurfactant behavior once they were placed in a media with unsaturated triglycerides as compared to those mixtures containing saturated triglycerides. Triglyceride agglomeration patterns, as well as triglyceride–protein interactions, appear to be dependent on the presence of double bonds within the fatty acid chains. While SASA values of OmpA and OmpN were similar, significant differences were observed in the SASA and β values when the system contains unsaturated triglycerides comparing with saturated ones. This result may be associated with the biosurfactant interaction with non-polar molecules. In conclusion, using an all atom MD, the authors accomplished a mechanistic understanding of protein interaction with molecules of different polarity. This approach could be the first step in a rational multiscale design of emulsions with particular emphasis on the cosmetic industry.

References

[1] M. Secchi, V. Castellani, E. Collina, N. Mirabella, and S. Sala, "Assessing eco-innovations in green chemistry: Life Cycle Assessment (LCA) of a cosmetic product with a bio-based ingredient," *J. Clean. Prod.*, vol. 129, pp. 269–281, 2016.

[2] N. Ranasalva, R. Sunil, and G. Poovarasan, "Importance of Biosurfactant in Food Industry," *IOSR J. Agric. Vet. Sci.*, vol. 7, no. 5, pp. 6–9, 2014.

[3] S. M. Aguilera-Segura, A. P. Macias, D. Carrero-Pinto, W. Lawrence, M.J. Vives-Florez, H.E. Castro-Barrera, O. Alvarez-Solano, A. F. Gonzalez-Barrios. "*Escherichia coli's* OmpA as Biosurfactant for Cosmetic Industry: Stability Analysis and Experimental Validation Based on Molecular Simulations," in *Advances in Computational Biology SE - 38*, vol. 232, L. F. Castillo, M. Cristancho, G. Isaza, A. Pinzón, and J. M. C. Rodríguez, Eds. Springer International Publishing, 2014, pp. 265–271.

[4] S. M. Aguilera-Segura, V. Nuñez-Vélez, L. Achenie, O. Alvarez-Solano, R. Torres, and A. F. González Barrios, "Peptides design based on transmembrane Escherichia coli's OmpA protein through molecular dynamics simulations in water–dodecane interfaces," *J. Mol. Graph. Model.*, 2016.

[5] H. Iwata and K. Shimada, *Formulas, Ingredients and Production of Cosmetics.* Saitama, Japan: Springer, 2013.

[6] S. Plimpton, "Fast Parallel Algorithms for Short–Range Molecular Dynamics," *J. Comput. Phys.*, vol. 117, no. 1, pp. 1–19, 1995.

[7] A. Stukowski, "Visualization and analysis of atomistic simulation data with OVITO–the Open Visualization Tool," *Model. Simul. Mater. Sci. Eng*, vol. 18, pp. 15012–7, 2010.

[8] W. Humphrey, A. Dalke, and K. Schulten, "VMD: Visual Molecular Dynamics."

532339 Prediction of Calcium Carbonate Wettability By Low Salinity Water Flooding Using Molecular Dynamics Simulations

Abstract:

Low salinity water (LSW) flooding has been considered by the oil industries in recent years as an augmented-water flood technique, which presents a substantial potential to improve the oil recovery factor. Although the LSW flooding technology was initially applied in sandstone reservoirs, the positive effects were also observed in those of the carbonate type. In sandstone reservoirs, low salinity water with less multivalent ions is usually employed, while sea water with more potential-determining ions such as Ca^{2+}, Mg^{2+} and $SO4^{2-}$ is generally injected into carbonate reservoirs. Rock wettability defines the tendency of the fluid (e.g. water) to spread or adsorb on the solid surface of rock grains in the presence of the other immiscible fluid (e.g. oil). In this work, we investigated the adsorption of an oil droplet mixture (13% Toluene : 87% Dodecane) on a calcite surface wetted with various numbers of water layers (1, 2 and 3 layers) containing different combinations of ionic species (NaCl, Na_2SO_4 and Na_2CO_3) as shown in figure 1. In order to determine the calcite wettability and spreading coefficients, we computed the surface/cohesive energies from MD simulations and fed the resulting parameters into the Young-DuprŽ equation.

The simulations show, in qualitative agreement with experiments, that the presence of water on the calcite surface strongly affects the adsorption of oil on calcite. Without water, oil is strongly adsorbed on the calcite surface and spreads into a thin layer, forming effectively a zero wetting angle. (This situation will reduce oil recovery.) A similar behavior is observed in the presence of a monolayer of water. Note that water is nominally hydrophilic, so one would expect that the oil should not spread. However, our previous work on the adhesion of naphthenic acid on calcite shows that the monolayer of water is strongly adsorbed on the surface of calcite, effectively making the surface hydrophobic (or conversely oleophilic). Thus we should expect oil to spread on the monolayer of water. But with more water molecules, we recover the expected behavior of oil to form a bead (or lens) on the surface of water.

532664 Thermodynamic Stability of Thiolate-Protected Gold Nanoclusters: From Molecular to Metallic Systems

Abstract:

Thiolate-protected atomically-precise gold nanoclusters (NCs) have generated strong interest due to their nonlinear optical properties, making them great candidates for applications in catalysis, chemical sensing, and energy conversion. Many gold NCs of distinct sizes have been successfully synthesized experimentally, yet there is still a lack of understanding when it comes to their stability. Two main questions that arise are why only certain size and compositions of gold NCs are stable and what role the ligands play in their stability. A promising approach to answer these questions is the newly developed thermodynamic stability theory[1], which revealed a parity between core cohesive energy (CE) and shell-to-core binding energy (BE) for every experimentally synthesizable nanocluster. Herein we use Density Functional Theory calculations to demonstrate that the theory can be extended to larger NCs, namely $Au_{146}(SR)_{57}$ and $Au_{279}(SR)_{84}$. Of the two NCs, the latter is of significance due to its metallic properties, which indicates that the theory can accurately predict stability independent of molecular or metallic NC behavior. The two NCs have also been found to match stoichiometric predictions between number of core metal atoms and number of ligands, thus demonstrating that one can accurately predict the exact stoichiometry of experimentally accessible NCs. We additionally address the effect of the type of ligands on the CE and BE contributions and on the overall thermodynamic stability of the NCs.

1. Taylor, M. G.; Mpourmpakis, G., Thermodynamic stability of ligand-protected metal nanoclusters. *Nat Commun* **2017,** *8*, 15988.

534881 Reactive Molecular Dynamics Simulation of Disintegration of Cross-Linked Epoxy-Resin Polymers upon Atomic Oxygen Bombardment

Abstract:

Ultraviolent radiation from the sun results in dissociation of molecular oxygen into atomic oxygen (AO) which is abundant in lower earth orbit. Re-entry vehicles entering earth atmosphere experience significant impact by these atomic oxygens, which cause both physical and chemical damage on the surface of the vehicle. The vehicle surface is normally made of polymer composites and evaluating these composites for AO impact requires an extensive experimental setup to simulate low earth orbit (LEO) conditions and is therefore expensive. Alternative to experiments, molecular dynamics is a powerful tool to examine the behavior of polymers with potential aerospace applications under harsh environmental conditions. In this study, we investigate the AO impact on various thermoset polymers using ReaxFF[1,2] reactive force field simulation. These cross-linked polymers are generated using our recently introduced accelerated cross-link methodology in ReaxFF. Since ReaxFF can simulate chemical reactions on-the-fly, both physical and chemical damages on polymers can be captured using ReaxFF simulations, which in turn, can be used in designing next generation polymer composites for aerospace application.

(1) van Duin, A. C. T.; Dasgupta, S.; Lorant, F.; Goddard, W. A. ReaxFF: A Reactive Force Field for Hydrocarbons. *J. Phys. Chem. A* **2001**, *105* (41), 9396â€"9409.

(2) Chenoweth, K.; van Duin, A. C. T.; Goddard, W. A. ReaxFF Reactive Force Field for Molecular Dynamics Simulations of Hydrocarbon Oxidation. *J. Phys. Chem. A* **2008**, *112* (5), 1040â€"1053.

519748 Molecular Simulation of CO_2 Absorption into MCM-41 Porous Material Filled with PDMS Solvent

Abstract:

Although CO_2 loadings into porous materials could be large, such as in MCM-41, the CO_2/N_2 loading selectivity in MCM-41 is small (~ 5 at 298 K), which is partly due to the large pore volume of MCM-41 available for N_2 adsorption. This low selectivity impedes the use of MCM-41 in CO_2/N_2 separation processes. In this work, we investigate a way to improve CO_2/N_2 loading selectivity by filling poly(dimethylsiloxane) PDMS solvent into the MCM-41 porous material. By using a smooth MCM-41 potential model, it was found that the simulated CO_2, N_2, and H_2 loadings into MCM-41 are close to the experimental data. At ambient condition, the PDMS solvent density filled in MCM-41 is 15% smaller than in the neat PDMS. Both CO_2 and N_2 loadings in MCM-41 filled with PDMS become smaller than their loadings in the empty MCM-41; however, it was found that the CO_2 loading decreases by 1.45 times and N_2 decreases by 3 times. These results lead to CO_2/N_2 loading selectivity in MCM-41 filled with PDMS solvent increasing by 2 times compared with CO_2/N_2 loading selectivity in the empty MCM-41. In the empty MCM-41, simulation snap shots show that at 298 K and 5 bar and 10 bar, 20% of the N_2 molecules are adsorbed in the regions away from the MCM-41 pore wall, while the remaining 80% of the N_2 molecules are adsorbed in the region close to the wall. For CO_2 at 298 K and 1.25 bar and 2.5 bar, only 7% of the CO_2 molecules are adsorbed in the region away from the wall and the remaining 93% of the CO_2 molecules are adsorbed close to the wall. These findings suggest that more N_2 molecules are adsorbed in the region away from the MCM-41 wall compared with CO_2. After PDMS is filled in the MCM-41, it will occupy the pores of the MCM-41 and will decrease gas loadings. On the other hand, simulation snap shots show that the PDMS in MCM-41 could also absorb both CO_2 and N_2 gases. The overall effect is that PDMS will decrease less CO_2 loading compared with N_2 loading and as a result increase the CO_2/N_2 loading selectivity. Finally, we will also show the H_2 adsorption in the MCM-41 filled with PDMS solvent.

529323 MD Simulation of a Magnesium Oxide Grain Boundary

Abstract:

Magnesium oxide is a ceramic compound with a melting temperature that makes it a desirable refractory material. It is also used as a protective coating on plasma displays, and as a diffusive barrier when treating radioactive waste. Diffusion through grain boundaries in similar materials is orders of magnitude faster than bulk diffusion, but is difficult to resolve experimentally.

We investigate how grain boundaries influence the transport properties of systems of MgO polycrystals consisting of parallel, layered MgO crystals. Simulations were run for a period of 100 ns. The systems were analyzed using an inherent structure approach, by removing thermal displacements from the system in post-processing. This allowed us to look for the unbalanced bonding at the grain boundary interface, and identify grain boundary atoms by their 6th nearest-neighbor distance. The mean squared displacement (MSD) was used to calculate self-diffusion coefficients of bulk and grain atoms. We demonstrate thermally induced grain mobility correlated with sliding and rotation of neighboring crystals. We also show the diffusivity as a function of distance from the center of the grain boundary, by obtaining diffusivities from MSD within planes normal to the boundary interface, to obtain an effective grain boundary width. We show atomic migration between the crystalline bulk and grain boundaries, and the effect of that migration on the effective grain boundary width. We show that magnesium self-diffusion is much faster than oxygen self-diffusion, and obtain values for the self-diffusion coefficient of magnesium.

537222 Atomistic Simulation of Sliding Friction between Two Silicon-Carbide Surfaces

Abstract:

Sliding friction between two SiC surfaces is important due to its applications to high performance disc brakes, cutting tools, and blades in gas turbines and jet engines, in addition to its traditional applications in the semiconductor industry. It is also important to study whether at the nanoscale the dependence of kinetic friction on sliding velocity follows Stokes' law. Since SiC exists both as an amorphous material and with a crystalline structure, the effect of surface roughness on the kinetic friction is also of importance. We report the results of extensive molecular dynamics simulation of sliding friction between surfaces of the two types of SiC over a wide range of sliding velocities. The amorphous SiC was generated by the reactive force field ReaxFF, which was also used for the friction simulations. The kinetic friction was computed by averaging the instantaneous friction force over sufficiently long distances. A normal force of 500 MPa was applied to the system, maintained at 300 K, and was used to determine the friction coefficient. At very low velocities, we find that Stoke's law is observed. At much higher velocities, however, there is a marked drop in the friction coefficient for both amorphous and crystalline SiC. For all the velocities, amorphous SiC exhibited a significantly higher friction coefficient. The nature of stick/slip conditions at varying velocities rests at the heart of this study, and has been investigated in detail.

536394 QSAR Modeling for Predicting Elimination Half-Life of Industrial Chemical Compounds

Abstract:

In recent years, there is an increasing interest for the development of Physiologically Based Toxicokinetic (PBTK) models, which provide quantitative descriptors of Absorption, Distribution, Metabolism and Excretion (ADME) of environmental or pharmaceutical chemicals. However, their application in toxicity testing and health risk assessment is limited due to the lack of input parameters required for their development. The proper parameterization of PBTK models is achieved using advanced Quantitative Structure-Activity Relationships (QSARs). QSARs are regression or classification models with many applications in toxicological sciences, aiming at modelling the biochemical interactions of chemical compounds with molecular targets and living tissues. A limited but very promising application of QSAR models, that are used to estimate the parameters of PBTK models, which provide descriptors of Absorption, Distribution, Metabolism and Excretion (ADME) of chemical compounds. The aim of this study is to develop QSARs for predicting the elimination half-life of industrial chemical compounds, as it is considered one of their major ADME properties. Moreover, elimination half-life is particularly important for PK models used in data-poor compounds, as well as for providing first-cut estimates of intake, starting from human biomonitoring data.

The methodological approach for the modeling of elimination half-life for environmental chemical compounds is presented. The first step of QSAR modeling was the preparation of input data, which included the existing experimental values of elimination half-life and the molecular descriptors of the corresponding chemical compounds. The input molecular descriptors were divided into two individual sets; the Linear Free Energy Relationship (LFER) and the PaDEL descriptors. The first set of descriptors included molar refractivity (E), solubility (S), acidity (A), basicity (B) and McGowan volume (V), which describe the LFER equation proposed by Abraham (1993). The second dataset consisted of 1444 1D and 2D descriptors, known as PaDEL descriptors. These descriptors are related to the molecular structure of the chemicals and are characterized as constitutional, topological, geometrical or electronic.

The dataset consisted of 199 industrial chemical compounds, which were randomly split to the training and the prediction set. The prereduction process was followed for the derived molecular descriptors in order to exclude the semi-constant and intercorrelated ones. Principal Component Analysis (PCA) was used for the distribution of chemical compounds. The score plot indicated that the molecules were clustered by structures, the loading plot showcased the most influential descriptors for the chemicals' categorization, while the screen plot gave the most significant principal components for data analysis. A genetic algorithm was then used for the selection of the optimal set of descriptors for the models.

The datasets were analyzed using two statistical methods, Multiple Linear Regression (MLR) and Artificial Neural Networks (ANN). MLR was implemented using the QSARINS software, while the ANN technique was implemented using the Neural Network Toolbox in Matlab. The Applicability Domain (AD) of the developed models was determined using the AD Toolbox in Matlab, based on several approaches such as bounding box, convex hull, leverage, distance to centroid, k Nearest Neighbors (kNN) approach and Probability Density Function (PDF)-based methods. The fitting performance (R^2) of the models using MLR and ANN was 0.70 and 0.87, respectively. The Leave One Out (LOO) technique (Q_{LOO}^2) indicated that model performance in predictions was equal to 0.65 and 0.81, while the external validation value (R_{ext}^2) was found to be 0.74 and 0.91 for MLR and ANN, respectively. The Mean Squared Error (MSE) for the training set ranged from 0.26 to 0.62, while for the prediction set ranged from 0.25 to 0.53. AD analysis showed that there were no outliers, verifying the reliability of each of the developed QSAR models. Elimination half-life was found to be greatly associated with plasma protein binding, as it is well known that chemical compounds with high protein binding tend to have greater values of elimination half-life. The derived descriptors define electrostatic, polarizability- and dispersion- type interactions, which strongly influence the affinity of molecules to the binding sites of plasma proteins and consequently their elimination half-life. Regarding LFER and PaDEL descriptors, it was proved that the second ones gave by far the best performance

Both models produced satisfactory results, with ANN, being the one with the best overall performance. The proposed models were checked for their fitting capacity, their validity and applicability. They were found to be stable, reliable and capable to predict physicochemical parameters of "data poor" chemical compounds that fall within the applicability domain that includes compounds with highly diverse properties such as VOCs, PBDEs, PCDDs και PCDFs. In this way, animal testing and laboratory experimentation could be reduced and the wide use of PBTK and PK models could be reinforced. Moreover, the "safe by design" concept for environmental chemicals is supported, by allowing the successful prediction of toxicokinetic behaviour based on molecular parameters, promoting thus green chemistry and cost saving of product development.

533028 Understanding Reaction and Transport in External Electric Fields with Molecular Simulations

Abstract:

External electric fields may directly affect the transfer of electrons and the movement of charged particles, Therefore, the control mechanism of external electric fields on chemical reaction and transport has been one of areas which receives lots of attentions. In this presentation, we will use molecular dynamics to investigate the control mechanism of external electric fields. The molecular simulations will base on classic or ReaxFF force fields. We will examine how electric fields will affect the growth of dendrites in Lithium metal batteries. We will also try to obtain molecular insights about how electric fields affect the oxidation of xylene. Experimental studies related to the simulations will also be discussed to verify simulation results.

537098 Electrophilic Aromatic Substitution and Intrinsic Nature of Aromaticity

Abstract:

Aromatic compounds are essential in the functioning of modern society and greatly affect the quality of everyday life. Aromatics-derived products are key to meeting the demands of pharmaceutical industry, transportation and packing industry, textile industry, electronics, and construction. Any fundamental breakthrough in the perception of the concept of aromaticity and the intrinsic nature of these compounds paves the way toward better understanding and more efficient utilization of this diverse class of molecules. Explaining the electronic structure of aromatic compounds, together with rational predictions regarding the reactivity and spectroscopic variables, provides a mean to define this venerable concept in a more strict manner. This will help us to understand the nature of these compounds, their reactivity and behavior to a higher extent. One of the interesting issues is the conceptual question whether the aromaticity represents the local or global property. In order to tackle this problem, we have examined the behavior of various aromatic compounds during the electrophilic aromatic substitution (EAS). In this work we considered how EAS affects the aromaticity in polycyclic aromatic compounds, conjugated systems (with the linkers of the variable length) and selected metallocenes. In order to fully understand these effects, we have used density functional theory, NICS value and other important aromaticity criterions. Finally, we have reproduced literature results regarding effects of benzene substitution on the system aromaticity, so we could more easily distinguish between effect originating from a different substituent and local/global aromaticity considerations. The obtained results and rationalizations establish a deeper insight into the nature of aromaticity and relate this understanding to their reactivity, thus providing a connection to their various applications.

537345 Prediction of Hg0 and $HgCl_2$ Adsorption Properties in UiO-66 Using Optimized Force-Fields

Abstract:

Predictions of Hg^0 and $HgCl_2$ Adsorption Properties in UiO-66 Using DFT-derived Force-Fields

Hongjian Tang,[†‡] Hanjun Fang,[‡] Yufeng Duan,[*,†] and David S. Sholl[*,‡]

[†]Key Laboratory of Energy Thermal Conversion and Control of Ministry of Education, School of Energy and Environment, Southeast University, Nanjing, Jiangsu 210096, China

[‡]School of Chemical & Biomolecular Engineering, Georgia Institute of Technology, Atlanta, Georgia 30332-0100, United States

Abstract: We propose the optimized force fields to effectively describe the vapor–liquid coexistence properties of Hg^0 and $HgCl_2$. For $HgCl_2$, an excellent agreement between theoretical and experimental data can be obtained using the Lennard-Jones potential. The pair potential of $HgCl_2$ predicted by LJ force field showed good consistence with that derived from CCSD(T). Since regular two-body pair potentials failed to reduplicate experimental density of Hg^0 in the bulk phase, Schwerdtfeger potential with multibody effect corrections was introduced to better describe the metallic interaction among Hg^0 atoms. For the sake of simulating $Hg^0/HgCl_2$ adsorption manners in UiO-66, generic force-field and mixing rules were applied to the GCMC calculation of $HgCl_2$ in UiO-66. While for Hg^0, a set of LJ force field parameters were obtained based on PBE-D3 method via random sampling and single-point energy calculations in the full range of accessible volume of void UiO-66 framework. Adsorption isotherms predicted by GCMC calculation showed that UiO-66 possessed an extremely high capacity for $HgCl_2$ adsorption but scarcely any loading of Hg^0 was observed in the framwork. UiO-66 was verified as a qualified material to selectively separate Hg^0 from $HgCl_2$ in industrial flue gas. Furthermore, the force field developed in the will be pioneering and promising in predicting adsorption behaviors of Hg^0 and $HgCl_2$ in MOFs and other porous materials through molecular simulation methods.

Key words: Hg^0; $HgCl_2$; Force field; Isotherms prediction; Selective separation;

526548 Using Artificial Neural Networks to Model Diffusion Characteristics in Lithium Solid State Electrolytes

Abstract:

There is great interest is developing solid state lithium electrolytes for use in an all solid-state battery to replace the flammable organic electrolyte. Previous efforts trying to understand the structure-function relationships resulting in high ionic conductivity materials have relied on *ab-initio* molecular dynamics (AIMD). Such simulations, however, are computationally demanding and cannot be applied to large systems containing more than a few hundred atoms in a reasonable time frame. Herein, we propose using machine learning artificial neural networks (ANN) to supply the forces and energies used during the MD simulations, and to eliminate the need of costly ab-initio force and energy evaluation methods, such as density functional theory (DFT). After carefully training a robust artificial neural network for four and five element systems, we obtain nearly identical lithium ion diffusivities for $Li_{10}GeP_2S_{12}$ (LGPS) when benchmarking the ANN-MD results with DFT-MD. We find that ANN-MD simulations allow the study of systems that require high number of atoms because the calculation of forces and energies scales linearly with the number of atoms rather than cubically in DFT (Figure 1), To demonstrate the power of the outlined ANN-MD approach we apply it to a chlorine doped LGPS system to calculate the effect of concentrations of chlorine on the lithium diffusivity at a resolution in $Li_{20}Si_3P_3S_{23}Cl$ with decreasing concentrations of chlorine: 2% 1% and 0.67%. Such low concentrations would be unrealistic to model with DFT-MD. We found that an optimal 1% chlorine substitution resulted in the highest predicted room temperature conductivity of 24.22 mS cm^{-1} which is in remarkable agreement with the experimental value of 25mS cm^{-1} (Figure 1).[1] We additionally test the overall robustness of these ANNs and find their accuracy is highly dependent upon the training data with some force fields being unable to reproduce 0 K structure optimization while others accurately reproduce both 0 K and finite temperature effects. Overall, we find that ANN-MD simulations can provide the framework to study systems that require a large number of atoms more efficiently and with high accuracy within the trained regime.

Figure 1: Plot of the time needed for one force calculation as a function of number of atoms. Above 50 atoms the artificial neural network is more efficient. In this region we calculate diffusion coefficients for a doped LGPS system and find an optimal diffusion at 1% Chlorine.

References:

1. Kato, Y. *et al.* High-power all-solid-state batteries using sulfide superionic conductors. *Nat. Energy* **1,** 16030 (2016).

531709 Using Free Energy Perturbation (FEP) to Rank Binding Affinities for ssDNA-Wrapped Single-Walled Carbon Nanotube (SWCNTs)

Abstract:

Separating and purifying single-walled carbon nanotubes (SWCNTs) is a challenge which currently inhibits their widespread application in fields that harness their unique electrical and optical properties. A novel strategy that has recently emerged for separating single-walled carbon nanotubes (SWCNTs) by chirality is the use of aqueous two-phase extraction (ATPE) in conjunction with single-stranded DNA (ssDNA) as a dispersant. This technique displays sequence-specific behavior which allows one to tune the separation in favor of particular SWCNT chiralities. The nature of this specificity is not well understood and optimal ssDNA/SWCNT pairs are currently searched by costly trial and error. In this talk, we describe a new protocol using molecular simulations and free energy perturbation (FEP) that can be used to quantify and rank the binding affinity of ssDNA-SWCNT pairs based on differences in their free energy of solvation. To test the method, we study the ssDNA-SWCNT binding process for both optimal and non-matching binding pairs and find that the ranking of ssDNA sequence binding affinity relative to specific SWCNT chiralities matches very well with experimental observations. We also study the effect of the relative concentration of ssDNA and use that information to quantify the effect of inter- and intra-strand ssDNA hybridization on the free-energy â€" yielding insight into the physics that drive the stability of the binding process. Future work to use the new protocol to design optimal binding pairs will also be described.

531114 Screening of Nano Porous Materials for Large-Scale Molecular Separations

Abstract:

A series of nano porous materials is investigated, with up to 1000 chemical compounds from various chemical families. About 400 MOFs from CoRE MOF database, and hypercrosslinked polymers with different percentage of STR, VBC, and DVB, were screened based on the affinity to the adsorbate and selectivity of each molecule pair. For some pseudomolecules in tested group, their physical properties such as critical temperature and critical pressure were estimated by using classical group contributions. We predicted the adsorption isotherm with a Langmuir isotherm model using the Henryâ€™s constant and saturation loading. Reasonable measurements exclude the cases with zero accessible void fraction or unphyscial heat of adsorption. Top performing nano porous materials were selected based on selectivity and working capacity.

528337 Computationally-Efficient High-Throughput Screening of Nanoporous Materials for Hydrogen Storage

Abstract:

Metal-organic frameworks (MOFs) are nanoporous materials consisting of metal nodes connected with organic linkers. These materials can have high surface area and porosity, which make them interesting candidates for applications in gas separation and storage. Since MOFs can be made from many different nodes, linkers, and functional groups, there are a huge number of possible MOF structures. High-throughput computational screening can efficiently characterize large numbers of materials to identify the best candidates for applications such as gas separation and storage. [1,2]

We have used high-throughput screening techniques paired with grand canonical Monte Carlo simulations (GCMC) to calculate the hydrogen storage capacity for over 130,000 MOFs. We will discuss commonalities in the structural and textual properties among the top-performing candidates from this screening and explore how the information gleaned from high-throughput screening can inform the design of new materials that are tailored for efficient hydrogen storage applications.

High-throughput screening generates a large amount of data, which makes it amenable to using machine learning and data mining to analyze the data for thousands of structures. We introduce an efficient regression model that can estimate the storage capacity for thousands of structures in a small fraction of the time required for full GCMC simulations. This allows us to quickly identify the most promising candidates for more detailed study, and this technique can be applied to other gases, such as methane, as well as a wide range of temperatures and pressures.

We demonstrate the utility of this screening method by searching a database of 55,000 known MOF structures [3] to identify the top candidates for hydrogen storage. After finding the best structures, we worked with experimental collaborators to synthesize one of the top structures and test its hydrogen storage capacity. This paradigm of using molecular simulation and machine learning techniques coupled with experiments can greatly accelerate the discovery and synthesis of advanced materials.

1. J. Colon, R.Q. Snurr, *Chem. Soc. Rev.*43, 5735-5749, 2014
2. S. Bobbitt, J. Chen, R. Q. Snurr, *J. Phys. Chem. C,*120, 27328-27341, 2016
3. Z. Moghadam, A. Li, S. B. Wiggin, A. Tao, A. G. P. Maloney, P. A. Wood, S. C. Ward, D. Fairen-Jimenez, *Chem. Mater.,* 29, 2618-2625, 2017

534689 Computational Design of Efficient Catalysts for CO_2 Hydrogenation

Abstract:

The atmospheric concentration of carbon dioxide (CO_2) has significantly increased over the last several decades, mainly caused by burning fossil fuels. The use of fossil fuels is likely to continue over the next several decades. A potential way to mitigate the increase of CO_2 in the atmosphere is carbon capture and recycling (CCR), where CO_2 is captured and combined with hydrogen obtained from renewable and nuclear energy sources to produce liquid fuels such as methanol. However, hydrogenation of CO_2 is very challenging due to its chemical inertness and thermodynamic stability. There is therefore a need to design efficient catalysts for CO_2 hydrogenation in a cost-effective way, such as computational simulation. This presentation will discuss three novel catalysts for CO_2 hydrogenation that covers heterogeneous catalysis, homogeneous catalysis and the ways to bridge the gap between both, which were computationally designed and investigated using density functional theory (DFT) combined with microkinetic modeling and multiconfiguration complete active space (MC-SCF) calculations. In the first part, I will briefly talk about indium oxide (In_2O_3). We have computationally predicted that In_2O_3 has good selectivity and stability for CO_2 hydrogenation to produce methanol. This prediction had recently been validated by experiments. Following that, I will present the mechanistic studies on bimetallic molecular catalyst for CO_2 hydrogenation in THF solvent at room temperature. Finally, I will focus on Lewis pair functionalized metal organic frameworks (MOFs), which is a material computationally designed by me, that has a low barrier pathway for CO_2 hydrogenation to produce methanol in this novel catalyst. Since MOFs have been shown to be useful for separating mixtures of gases, MOFs functionalized with optimal catalytic functional groups, such as Lewis pairs, allow us to combine CO_2 capture and conversion in a single material, which could greatly improve the economics through process intensification.

538406 Quantum Mechanics Based Multiscale Reactive Simulations of Materials and Processes

Abstract:

There is an economic and environmental imperative to develop new materials. Critical to the development and manufacturing of new generations of materials is the development of first-principles based multiscale multiparadigm methods and modeling software scalable from atoms to continuum, capable of efficiently predicting the dynamical behavior (reactive dynamics and kinetics) of the homogenous and heterogeneous materials. These methods are suitable for significantly improved performance of next generation materials. Recent dramatic developments in first-principles quantum mechanics (QM) methods has enabled accurate predictions for the properties of different systems. But the QM methods are limited to ~200 atoms and time scales of ~10 picoseconds (ps). In contrast, the developments of the new materials needed for dramatic improvements in the properties of materials under extreme conditions and electrochemical systems, for example, require spatial scales of 100 nm and beyond (>100 million atoms) and time scales of microseconds and beyond. Thus, there is an enormous gap between the scale of current QM methods and that of the real applications.

In my research I **develop and validate the computation tools and software** to fill this gap while simultaneously applying these methods to developing new generations of materials. Of my particular interest is the applications of these methods to electrochemical processes: **batteries**, **fuel cells**, and materials under extreme conditions: **high energy density matters**.

In this talk, I provide the details of development of these new advanced methods that provide near QM accuracy for reactive simulation of large systems. This provides the basis for *in silico* discovery of new combinations of materials that can subsequently be optimized.

In particular, I discuss the results of the simulations for water system because of its central role in life and importance in a lot of applications. We find quite excellent agreement with experimental data for solid and liquid phase of water: T_{melt}=273.3K (exp=273.15K) and properties at 298K: Î"H_{vap}=10.36 kcal/mol, density = 0.997 g/cc, entropy= 68.4 eu, dielectric constant=76.1, ln Ds (self-diffusion coef) =-10.08 compared to experimental values of 10.52, 0.997, 69.9, 78.4, and -11.24, respectively. We expect this model to remain accurate as a function of temperature and pressure. We have used this force field to study the properties of water at the surface including surface thickness, water orientation, hydrogen bond distribution, and vibrational frequencies which are experimentally hard to obtain. In addition, we have discovered for the first time the existence of six-coordinated water molecules at the areas close to the surface which are stable over 10 ps time intervals which could be responsible for some of the complicated water properties at the surface. I expect that we can use the above methods to extend it to ions, proteins, DNA, polymers, and inorganic systems for applications to biomolecular, pharma, electrocatalysis (fuel cells, water splitting) and batteries where interactions with explicit water molecules plays a significant role.

538652 Reaction Ensemble Monte Carlo Simulations of Protic Ionic Liquid Formation

Abstract:

A protic ionic liquid (PIL) is formed when an acid and base exhibit a high degree of ionization upon mixing

$$AH + B \rightarrow [A^{-}]\,[BH^{+}]$$

and the mixture, by convention, has a melting point below 100 °C. The ionization of an acid-base pair is difficult to predict. The difference between the aqueous pK_a of the acid and that of the base has shown some correlation with the boiling point of the mixture [1]. However, when paired with the same acid, bases with similar pK_as may exhibit very different ionic conductivities [2].

Reaction ensemble Monte Carlo is one promising method to predict which acid-base pairs form PILs. The challenge lies in developing Monte Carlo moves that efficiently insert and delete charged species into an ionic liquid environment. Simple insertions, semi-grand insertions [3], and state-of-the-art continuous fraction component (CFC) insertions all fail at such a task. We present a combination semi-grand/CFC move that succeeds and demonstrate its effectiveness on ethylammonium nitrate and a series of bases paired with acetic acid.

This work performed under the auspices of the U.S. Department of Energy by Lawrence Livermore National Laboratory under Contract DE-AC52-07NA27344. Document release number LLNL-ABS-749853.

References

1. Yoshizawa, W. Xu and C. A. Angell, "Ionic Liquids by Proton Transfer: Vapor Pressure, Conductivity, and the Relevance of ΔpK_a from Aqueous Solutions", *J. Am. Chem. Soc.*, **2003**, 15411-15419
2. Stolmenovskl, E. I. Izgorodina and D. R. MacFarlane, "Ionlclty and Proton Transfer In Protlc Ionlc Liquids", *Phys. Chem. Chem. Phys.*, **2010**, 12, 10341-10347
3. G. Mullen and E. J. Maginn, "Reaction Ensemble Monte Carlo Simulations of Xylene Isomerization Under Confinement", *J. Chem. Theory Comput.*, **2017**, 13, 4054-4062.

531225 Estimation of Nucleation Barriers for Colloidal Crystals from Computer Simulations

Abstract:

One of the most prevalent phase transformations in nature is the nucleation of crystals from fluid phases and their growth. Applications range from ice crystal formation in the atmosphere, to metallurgy, nanomaterials, and protein crystallization. Despite its overwhelming importance, the details of crystal nucleation remain poorly understood. In classical nucleation theory, the barrier of homogeneous nucleation is given by two contributions, the energy gain of creating a nucleus and the energy loss due to surface tension of the new created interface. While nuclei are in general non spherical and have anisotropic interfacial tension, the classical nucleation theory neglects those effects.

In this talk, we present a method to calculate nucleation barriers from Monte Carlo simulations without the need to precisely locate the interface or to compute the anisotropic interfacial tension [Statt, Antonia, Peter Virnau, and Kurt Binder. *Phys. Rev. Lett.* 114.2, 026101 (2015)]. As a test case, we investigate a soft extension of the well-known effective Asakuraâ˘'Oosawa model for colloidâ˘'polymer mixtures [Dijkstra, Marjolein, RenÃ© van Roij, and Robert Evans. *Phys. Rev. E* 59.5, 5744 (1999)]. While our analysis is appropriate for crystal nuclei of arbitrary shape, we find the nucleation barrier to be compatible with a spherical shape and mostly consistent with classical nucleation theory.

525212 Replicating the Static and Dynamic Behavior of a Hybrid Fluid Via Relative Resolution

Abstract:

Recently, we presented a novel framework for multiscale simulations, termed "Relative Resolution" (RelRes) [1]. Importantly, our algorithm contains both Fine-Grained (FG) and Coarse-Grained (CG) models in a single fluid system. The unique feature of RelRes is that it switches between molecular resolution in terms of relative separation: While nearest neighbors are characterized by a FG (geometrically detailed) model, other neighbors are characterized by a CG (isotropically simplified) model. Instead of performing an iterative procedure for relating the FG and CG models [2], we formulate for this task an analytical expression which is based on a multipole approximation at appropriate distances: Stemming in a Taylor expansion that conserves energy between a molecular pair in an infinite limit, our algebraic equation allows for parametrization between FG and CG models on a paper sheet instead of a computer cluster. Except that our formalism connects with classical theories in statistical mechanics [3], we also show that RelRes is the natural version of the cell-multipole algorithm for molecular systems [4]: Specifically, our approach groups sites in discrete molecules rather than in continuous cells. With this hybrid formalism for fluid mixtures, we consequently study several systems in molecular simulations (e.g., a liquid-liquid of oxygens-nitrogens, a liquid-gas of tetrachloromethanes-thiophenes, etc.), examining for them both their static and dynamic behavior. We also look at the effect of changing the ratio between the number of FG and CG sites in a given oligomer. As a next step, we proceed by implementing RelRes for water. In summary, RelRes can enhance the computational efficiency of molecular simulations by orders of magnitude, while still capturing the correct phase diagram of various mixtures.

[1] A. Chaimovich, C. Peter, and K. Kremer. Relative resolution: A hybrid formalism for fluid mixtures. The Journal of Chemical Physics 143:243107, 2015.
[2] M. G. Saunders and G. A. Voth. Coarse-graining methods for computational biology. Annual Review of Biophysics 42:73-93, 2013.
[3] B. Widom. Intermolecular forces and the nature of the liquid state. Science 157:375-382, 1967.
[4] L. Greencard and V. Rokhlin. A fast algorithm for particle simulations. Journal of Computational Physics 73:315-348, 1987.

537439 Using Advanced Field-Based Approaches to Predict Polymer Nanocomposite Phase Behavior

Abstract:

The considerable parameter space associated with polymer nanocomposites (PNCs) requires efficient theoretical and simulation methods to understand their assembly and guide experiments. Typical PNC modeling approaches are limited by their computational expense or use of critical approximations. Here we describe the development of field-based approaches in modeling PNCs and demonstrate their efficacy in efficiently modeling a large parameter space with few approximations. The exciting advances of these field-based approaches have led to a description of large scale phase behavior, including phase diagrams, for an array of polymer nanocomposite systems. These systems include grafted nanoparticles in a polymer melt and the assembly of mixed brush nanoparticles in solution. In particular, we find the introduction of thermal fluctuations and non-uniform grafting sites on the surface of the nanoparticles have a significant impact on the overall phase behavior of these systems. Finally, we further show the versatility of these methods to other PNC systems and provide comparisons with experimental results.

536489 A Graph-Based Approach for Systematic Molecular Coarse-Graining

Abstract:

An overarching aspiration in soft-materials research is the theory-based design of systems with desired functional properties. Some examples include identifying polymer chemistries and architectures that facilitate fast ion conduction for use in batteries or synthesizing polymers with requisite solubility and interactions that enable supra-molecular assembly for application as biomaterials. While computational statistical mechanics provides a framework to establish structure-property relationships that facilitate design, many important phenomena require descriptions over a broad range time and length scales that extend beyond the capabilities of atomistic molecular dynamics and require alternative simulation strategies. At the same time, these alternative simulation strategies should be generalizable and automatable for purposes of design or materials characterization.

In this talk, I will discuss a novel coarse-graining approach that enables efficient simulation of molecular systems at a targeted resolution. A central feature of this coarse-graining approach is a systematic, graph-theoretic coarse-graining scheme, which provides an unambiguous and automated way to generate coarse-grained representations that preserve chemical topology and reflect chemical functionalization. I first illustrate the approach using a series of simple systems before application to a series of high-molecular weight polysaccharides, for which atomistic simulations are infeasible. Using a series of models with increasing levels of â€œcoarseness,â€ I characterize relevant structural and dynamic properties and assess how the choice of coarse-grained representation affects such predictions at different time and length scales. Overall, these studies provide valuable mechanistic insight, and the described methodology provides a framework with significant utility for the coarse-grained study of various molecular systems.

532400 Data-Driven Modeling in Molecular Science and Chemical Engineering

Abstract:

Recent advances in data-driven modeling and deep machine learning have revolutionized the fields of computer vision, natural language processing, and game playing. Inspired by these recent success, I demonstrate the application of several methods from deep generative modeling and reinforcement learning on problems in molecular design and discovery, transport phenomena, and process control.

532639 Modeling Heterogeneous Electrocatalysis on Realistic Surfaces from First-Principles, Thermodynamics, and Machine Learning

Abstract:

Theoretical and computational approaches, most notably the *d*-band model for chemisorption scaling relations, have made important contributions to the understanding of heterogeneous electrocatalysis. Such approaches, however, implicitly assume that the catalytically-active surface for a particular class of materials is the same. We have discovered that the electrocatalytic activity of nickel phosphides (Ni_2P, Ni_5P_4, and Ni_3P) and calcium manganate ($CaMnO_3$) toward the hydrogen and oxygen evolution reactions, respectively, are governed by aqueous surface equilibria, *i.e.* the most catalytically-active surfaces are aqueous surface reconstructions. Nickel phosphide surfaces react with phosphoric acid at small, reducing electrode potentials to form P-enriched surface reconstructions that offer nearly thermoneutral H adsorption. Conversely, for calcium manganate, it is Mn-depleted surface reconstructions, achieved via the formation of surface Mn vacancies, that activate lattice oxygen and lower the overpotential for oxygen evolution. In addition to the determination of realistic surfaces for computational studies of heterogeneous electrocatalysis, we are also developing combined first-principles and machine learning techniques for the automated discovery of catalytic descriptors. Using these techniques, we discovered that the Ni-Ni bond length is an excellent descriptor for the hydrogen evolving activity of Ni_2P and that it can be modulated via nonmetal surface doping, which induces a chemical pressure effect.

527092 Uncovering Reaction Maps to Promote Active Catalysis

Abstract:

Complete knowledge of reaction mechanisms allows catalysts to be designed and operated at high activity, using minimal empiricism and guesswork. Learning the mechanism of catalysis, however, is a challenging task that is usually guided by substantial prior intuitions and analogies. To compound this difficulty, most mechanistic knowledge is centered on active catalytic cycles, and little information is available for side reactions which may hinder or even prohibit catalysis. In this presentation, computational methods will be introduced for exploring mechanisms in reactions catalyzed by transition metal complexes, with a special focus on off-cycle pathways and reactive steps which cripple the desired activity. A description of these new tools will be given alongside thoughts about future developments and incorporation of machine learning into the reaction prediction landscape. The focus applications of our tools will include C-C bond forming reactions catalyzed by group 10 metals, for instance the construction of advanced polymeric materials for solar cell applications.

528366 Liquid Phase Modeling in Heterogeneous Catalysis

Abstract:

The role(s) of solvent in affecting the intrinsic activity and selectivity of active sites, as well as having a significant impact on the reaction mechanism of heterogeneously catalyzed reactions is well-recognized nowadays. Liquid phase processing is attractive for the conversion of highly functionalized lignocellulosic biomass, considering the aqueous condition of the feedstock and their high-water solubility, reactivity, and thermal instability. However, the ability to include the effect of a liquid phase environment on chemical reactions occurring at solid-liquid interfaces poses a distinctive challenge to computational catalysis. Describing such a system from first principles is computationally very demanding, often requiring configuration space sampling of an aqueous phase environment consisting of hundreds if not thousands of molecules.

The main objective of our study is to gain a better understanding of the chemistry of heterogeneously catalyzed reactions at solid-liquid interfaces using both implicit and explicit solvation approaches. As a prototypical reaction with relevance to biomass catalysis, we chose to investigate the hydrodeoxygenation (HDO) of ethylene glycol over Pt(111) and Ru(0001) under aqueous phase processing conditions by employing both our iSMS[1] (Implicit Solvation for Metal Surfaces) and eSMS[2] (Explicit Solvation for Metal Surfaces) solvation models. Our results indicate that implicit solvation models are inherently restricted by their inability to characterize strong directional hydrogen bonding while the hybrid QM/MM models can predict reasonably accurate solvent effects with an adequate balance between computational expense and chemical accuracy of the reaction system.[3] To describe the metal-water interaction at DFT level accuracy, we developed a high dimensional neural network potential using supervised machine learning techniques for representing the interaction between Ru(0001) and water. Mapping a high dimensional continuous potential energy surface allows us to calculate energies and forces for an arbitrary metal-water structure[4] and thus, this neural network potential can be used in our hybrid QM/MM solvation formalism for modeling chemical processes at Ru-water interfaces.

(1) Faheem, M.; Suthirakun, S.; Heyden, A. *J. Phys. Chem. C* **2012**, *116*, 22458-22462.

(2) Faheem, M.; Heyden, A. *J. Chem. Theory Comput.* **2014**, *10*, 3354-3368.

(3) Saleheen, M.; Heyden, A. *ACS Catalysis* **2018**, *8*, 2188-2194.

(4) Behler, J. *J Phys-Condens Mat* **2014**, *26*, 183001:183001-183024.

531605 Improving the Initial Guess for a Nudged Elastic Band Calculation By Incorporating Chemical Intuition

Abstract:

The Nudged Elastic Band (NEB) is a powerful technique incorporated into multiple computational chemistry packages for the elucidation of complex reaction pathways [1, 2]. Although modern computers are certainly capable of handling these calculations, their execution, as with all iterative mathematical programs, can be severely affected by the initial guesses provided by the user. Typically, the initial guesses for reaction pathways are constructed via linearly interpolating the coordinates of atoms from an initial state to a final state. Recently, Smidstrup et al. [3] proposed a method where bond distances between atoms were interpolated rather than coordinates and showed performance improvements to the respective NEB calculations using these improved initial guesses. However, as most calculations using the NEB method are for systems with more than seven atoms, bond distances cannot be perfectly interpolated along a reaction pathway due to a lack of degrees of freedom in the involved optimizations (more bonds than atomic coordinates to move). To this end, Smidstrup et al. weighted the bonds in their construction of initial guesses based on the inverse quartic bond length, which ensured that shorter were interpolated at the expense of longer ones.

As there may certainly be cases where shorter bonds may not require accurate interpolation and longer ones might, we chose to turn to a userâ€™s chemical intuition to guide these improved interpolations. Our approach allows the user to specify which bonds, based on their knowledge of the underlying chemistry, should be considered important in the construction of an initial guess for a given reaction pathway, which are implemented in an objective function for an NEB-style minimization of the reaction pathway. Bonds not selected to be important are ignored in the aforementioned objective function and therefore the associated atoms only move in accordance to the artificial spring forces present during the NEB minimization. Further, the specific computational implementation allows for the user to specify how they would like bonds to change throughout the reaction pathway (not strictly limited to pure linear interpolation). Here, we present the performance of our method for constructing initial guesses to that of Smidstrup and strict coordinate linear interpolation for a variety of reaction systems through analysis of the ionic steps required for the subsequent quantum mechanical NEB calculations. We also present a specific case where initial guesses generated through our approach proved able to adequately map a reaction pathway while coordinate linear interpolation for an initial guess gave unrealistic results [4].

References:

[1] G. Henkelman, H. Jónsson, Improved tangent estimate in the nudged elastic band method for finding minimum energy paths and saddle points, The Journal of Chemical Physics, 113 (2000) 9978-9985.

[2] G. Henkelman, B.P. Uberuaga, H. Jónsson, A climbing image nudged elastic band method for finding saddle points and minimum energy paths, The Journal of Chemical Physics, 113 (2000) 9901-9904.

[3] S. Smidstrup, A. Pedersen, K. Stokbro, H. Jónsson, Improved initial guess for minimum energy path calculations,

The Journal of Chemical Physics, 140 (2014) 214106.

[4] A.J.R. Hensley, J. Zhang, I. Vinçon, X.P. Hernandez, D. Tranca, G. Seifert, J.-S. McEwen, Y. Wang, Mechanistic understanding of methanol carbonylation: Interfacing homogeneous and heterogeneous catalysis via carbon supported Ir-La, Journal of Catalysis, 361 (2018) 414-422.

533515 Improving the Efficiency of Kinetic Monte Carlo Simulations for Catalysis with a Parallel Caching Algorithm

Abstract:

Improving the Efficiency of Kinetic Monte Carlo Simulations for Catalysis with a Parallel Caching Algorithm

Kinetic modelling methods have proven important for the fundamental understanding of underlying mechanisms that give rise to catalysis, but also for the prediction of catalytic performance metrics, such as activity, selectivity and stability (e.g. poisoning resistance). Kinetic Monte Carlo (KMC) approaches have shown exceptional promise in tackling complexities relevant to the effect of multiple adsorbate configurations, lateral interactions, adlayer structure, and the presence of spectator species, to name a few factors, on activity. The main issue with KMC simulations is that they are computationally expensive.

In this talk, we will discuss an efficient implementation of the graph theoretical KMC framework[1] that employs a caching scheme with OpenMP parallelisation to improve computational performance. This KMC framework incorporates Brønsted-Evans-Polanyi (linear) relations to calculate an activation barrier from the reaction energy, the latter evaluated as the difference in the initial and final state species of a given elementary event, in the presence of interactions with spectators. Thus, after a reaction has occurred, the removal of reactants and addition of products can affect the barriers (and thus rates) of other reactions in the neighbourhood of the one that just happened. The key concept behind the caching scheme is that, in order to update the barriers of the "affected reactions" one only needs to detect interactions between the product species of the "just-executed" reaction and the products of the "affected reactions". Remaining interaction patterns, between the products of the "affected reactions" and other spectators are cached in a datastructure. Benchmarks on a model of the NO oxidation/NO_2 reduction reaction on Pt, show that the parallel caching scheme can exhibit up to 20x improvement in efficiency compared to a single processor run.

References

1. Nielsen, W., d'Avezac, M., Hetherington, J. and Stamatakis (2013). "Parallel kinetic Monte Carlo simulation framework incorporating accurate models of adsorbate lateral interactions". *The Journal of Chemical Physics*, 139(22): 224706.

537800 Machine Learning Molecular Dynamics for Understanding Nonadiabatic Surface Reactions

Abstract:

Transferring energy from solid surfaces to chemical bonds of adsorbed species is a fundamental process in heterogeneous catalysis. To a great extent, the dynamics of molecule-surface interactions driven by thermal energy, i.e., heat, can be well described within the framework of the Born-Oppenheimer approximation. New reaction channels[1] often open up in response to electronic excitations via surface-mediated energy transfer. In this process, the energetic electrons or holes can scatter into surface species and heat up vibrational modes of adsorbates. This process involves not only the electronic ground state but also the excited state, and it is nonadiabatic with respect to the nuclear motions[2].

In the weak coupling limit, the nonadiabatic energy transfer on metal surfaces can be modeled using only the ground-state potential energy surfaces and the nonadiabacity is taken into account with electronic friction contributions in Langevin dynamics by solving Newtonian equations with a thermal fluctuation term[2,3]. However, it is extremely time consuming to compute electron-phonon coupling strength for large systems.

In this talk, we will discuss our most recent development of a machine learning molecular dynamics approach that prorogates the system by 'learning from data'. Machine learning algorithms, such as the artificial neural networks[4], can use past trajectories as training datasets for fast and accurate prediction of interatomic forces and friction force, thus allows us to perform statistical analysis of nonadiabatic surface reactions. We benchmark the approach for activation of oxygen from hollow to bridge and CO oxidation on Ru(0001) because of strong interests in understanding laser-induced chemistry on metal surfaces. Local Density Friction Approximation (LDFA) was used to calculate the electronic friction. Statistical analysis from ~5000 molecular dynamics trajectories shows the oxygen hopping under ultrafast laser pulse is electron driven in nature.

[1] M. Bonn *et al.*, "Phonon- Versus Electron-Mediated Desorption and Oxidation of CO on Ru(0001)," *Science*, vol. 285, no. 5430, pp. 1042–1045, Aug. 1999.

[2] M. Head Gordon and J. C. Tully, "Molecular dynamics with electronic frictions," *J. Chem. Phys.*, vol. 103, no. 23, pp. 10137–10145, Dec. 1995.

[3] M. Brandbyge, P. Hedegård, T. F. Heinz, J. A. Misewich, and D. M. Newns, "Electronically driven adsorbate excitation mechanism in femtosecond-pulse laser desorption," *Phys. Rev. B*, vol. 52, no. 8, p. 6042, 1995.

[4] A. Khorshidi and A. A. Peterson, "Amp: A modular approach to machine learning in atomistic simulations," *Comput. Phys. Commun.*, vol. 207, pp. 310–324, Oct. 2016.

538818 Mapping the Degree of Rate Control Using Automated Construction of Microkinetic Models with Rmg-Cat

Abstract:

Kinetic parameters for surface reactions can be predicted using a combination of DFT calculations, scaling relations, and machine learning algorithms; however, construction of microkinetic models still requires a knowledge of all the possible, or at least reasonable, reaction pathways. Our recently developed Reaction Mechanism Generator for Heterogeneous Catalysis (RMG-Cat) [1], built upon the well-established open-source RMG software [2], can provide the complete reaction network from user-supplied initial conditions. It can: propose both elementary reactions and new species by applying reaction templates and recipes; estimate thermodynamic and kinetic parameters using databases of DFT calculations, functional-group decision trees, and heuristics; solve governing rate equations; and decide which reaction pathways should be explored further.

Using linear scaling relations RMG-Cat can now estimate adsorbate thermochemistry on a range of hypothetical metal surfaces and construct detailed microkinetic models for any metal surface, with no prior assumption of a rate limiting step. By running simulations with sensitivity analysis, we can now determine the rate limiting step on each surface and can plot a “volcano surface” of the degree of rate control of each reaction as a function of elemental binding energies. We demonstrate this new tool on the catalytic combustion of methane.

[1] Goldsmith, C. F., West, R. H. “Automatic Generation of Microkinetic Mechanisms for Heterogeneous Catalysis” J. Phys. Chem. C. 21 (18), 9970–9981 (2017) https://doi.org/10.1021/acs.jpcc.7b02133
[2] Gao, C. W., Allen, J. W., Green, W. H., West, R. H. "Reaction Mechanism Generator: Automatic construction of chemical kinetic mechanisms". Comput. Phys. Commun., 203, 212-225, (2016) https://doi.org/10.1016/j.cpc.2016.02.013

536428 Modelling of the Dynamic Behavior of Catalyst Materials in Reacting Conditions: An Application to the Catalytic Partial Oxidation of Methane on Rhodium

Abstract:

Modelling of the dynamic behavior of catalyst materials in reacting conditions: an application to the catalytic partial oxidation of methane on rhodium

Raffaele Cheula[1], Aloysius Soon[2], Matteo Maestri[1*]

[1]Laboratory of Catalysis and Catalytic Processes, Dipartimento di Energia, Politecnico di Milano, via La Masa 34, Milano, Italy

[2]Department of Materials Science and Engineering, Yonsei University, 50 Yonsei-ro Seodaemun-gu, Seoul, Republic of Korea

[*]Corresponding author: matteo.maestri@polimi.it

1. Introduction

Several experimental studies demonstrated that the activity of heterogeneous catalysts is strongly dependent on their structure [1]. Moreover, the catalyst structure is deeply influenced by the reaction environment. As a result, catalyst materials in heterogeneous catalytic reactors are intrinsically dynamic systems: they change their size and shape in response to the conditions of the reaction environment, which in turn affects the reactivity [2]. Microkinetic modelling is a key tool for the investigation of the behavior of complex chemical reactions, making it suitable for the fundamental understanding of the structure-activity relations of catalytic processes. However, state-of-the-art microkinetic models in heterogeneous catalysis lack in the description of the catalyst structure. In fact, they typically rely on an abstract and "structureless" concept of catalyst active site. This approach leads to a satisfactory description of conversion and selectivity trends and therefore it is a very valuable tool for the description of the macroscopic kinetic behavior of a catalytic process [3]. However, the material gap in the modelling of the active sites intrinsically precludes the understanding of the underlying mechanisms at the atomic-scale level. As a consequence, the incorporation of the catalyst structure in microkinetic modelling becomes of paramount importance for the fundamental understanding of the structure-activity relations, which is widely recognized as one of the main progress area in modelling of catalysis towards the design and the optimization of catalysts based on functional understanding rather empirical testing. In this contribution we couple microkinetic modelling and *ab initio* thermodynamics for the characterization of catalyst nanoparticles in terms of surfaces morphology and three-dimensional shape as a function of the reaction environment. This allows to calculate the distribution of the catalyst active sites and how it changes during reaction. The methodology is applied in the context of the CH_4 catalytic partial oxidation and reverse water-gas shift reactions on Rh/Al_2O_3.

2. Methods

Multiscale reactor modelling is employed for the calculation of the gaseous composition profiles inside a chemical reactor and for the identification of the most abundant reaction intermediates (MARIs) present on the catalyst surfaces. Density functional theory calculations and *ab initio* thermodynamics are then exploited to calculate the most stable bulk and surface structures of the catalyst at different conditions of the reaction environment. With this approach we consider the presence of the MARIs in thermodynamic equilibrium with their reservoirs in the gas phase surrounding the catalyst. Atomistic Wulff-Kaishew construction is then applied to calculate the three-dimensional shape of the supported catalyst nanoparticles as function of the chemical environment inside the reactor. Given the particles size distribution from HRTEM images, we accurately calculate the catalyst dispersion and the distribution of the different types of the catalyst active sites and how it changes during reaction.

3. Results and discussion

The multiscale reactor modelling of the CPO on Rh/Al_2O_3 reacting system shows a sharp change in the reaction environment with residence time and temperature. In particular, when oxygen is present in the boundary layer around the catalyst surfaces, the production of CO and H_2 is not observed (zone 1 in Fig. 1a). Once oxygen has been totally depleted, syngas starts producing (zone 2 in Fig. 1a). We investigate the most stable structure of the catalyst in these two different zones of the reactors. First, the most stable bulk phase of the catalyst is calculated, considering the thermodynamic equilibrium with the gas phase. Then, the morphology of different crystal facets of the nanoparticles is analyzed: for each facet, the thermodynamically most stable surface structure in equilibrium with the surrounding gas is calculated. The results are reported in terms of bulk phase diagram in Figure 1b as a function of temperature and partial pressure of O_2. We find that the change from zone 1 to zone 2 is accompanied by a change in the bulk structure and the morphology of the catalyst. In the oxidizing zone of the reactor (zone 1), Rh_2O_3 is the thermodynamically stable bulk phase, and the nanoparticle exposes mainly the (0001), (1-102) and (11-23) crystal facets (Fig. 1b). In zone 2 of the reactor, instead, where O_2 drops to very low values at the catalyst interface, metallic Rh becomes the most stable bulk structure. CO* and H* are the MARIs, and Rh(100) and the Rh(111) are the facets most exposed by the nanoparticle. High Miller index facets – in particular Rh(311) and Rh(331) – become more stable with the increase of coverage of the MARIs, observed during reaction, yielding an important change in shape of the catalyst and in the distribution of the active sites. The change of the oxidation state of Rh as a response of the change in O_2 concentration in the gas phase is in agreement with operando spectroscopy studies [3]. Given the particle size distribution from HRTEM images, we calculate the dispersion of the catalyst by filling Wulff-Kaishew plots with Rh atoms and counting the resulting number of atoms that constitute the catalyst surface. Good agreement is found with experimental values of CO and H_2 chemisorption. The framework is applied also to the reverse water-gas shift reaction on Rh/Al_2O_3. The distribution of the active sites is found to change with the amount of CO in the reacting environment. The three-dimensional map of the active sites of the nanoparticles, is then translated to a two-dimensional grid for kinetic Monte Carlo simulations.

4. Conclusions

Multiscale reactor modelling, *ab initio* thermodynamics and Wulff-Kaishew construction have been coupled for predicting the morphology of supported catalyst nanoparticles and its variation under reacting conditions. Our results – in agreement with spectroscopic studies – show that Rh can undergo strong variation of bulk phase composition and in the active sites distribution in response to different reaction environment found in the reactor.

Acknowledgments

Financial support from the European Research Council is gratefully acknowledged (ERC project SHAPE, grant n° 677423).

Figure 1. Major species partial pressures axial profiles at the catalyst interface. T = 500°C. P = 1 atm. Inlet mole fractions: CH_4 = 0.01, O_2 = 0.01, N_2 = 0.98. Gas hourly space velocity: 2E6 $Nl/kg_{cat}/h$. Annular reactor: I.D. 4 mm, O.D. 5 mm, Length 22 mm [2]. (b): Bulk phase diagram of Rh (blue) and Rh_2O_3 (red). (c): Atomic structure of a nanoparticle calculated by Wulff-Kaishew construction, size taken from the particle size distribution obtained by HRTEM images. (d) Catalyst active sites 3D grid.

References

[1] G. Ertl: Angew. Chem. Int. Ed., 47 (2008) 3524-3535
[2] M. Maestri. D. G. Vlachos, A. Beretta, G. Groppi, E. Tronconi: Top. Catal., 52 (2009) 1983-1988
[3] J. D. Grunwaldt, S. Hannemann, C. G. Schroer, and A. Baiker: J. Phys. Chem. B, 110 (2006) 8674-8680

530364 Effects of Dopant Loading and CO Adsorption on the Structural Stability of Highly Dilute Alloys

Abstract:

Platinum group metals (PGMs), including Pd, Pt, Rh and Ir, as well as Ni are often employed to catalyse reactions of high industrial importance owing to their excellent activity in a wide range of applications. Yet, their high catalytic activity is often hampered by their susceptibility to CO poisoning that in turn leads to catalyst deactivation.[1,2] Studies have shown that this problem may be effectively addressed by preparing highly dilute alloys of PGMs with less reactive host metals, such as Ag, Au and Cu. In particular, single atom alloys (SAAs) are found to exhibit high tolerance to CO poisoning as a result of the weaker CO binding compared to PGMs,[3] whilst offering the economic benefit of minimizing the loading of expensive PGMs. These advantages in conjunction with their excellent catalytic performance, in terms of activity and selectivity in a variety of reactions such as hydrogenations,[4] dehydrogenations[5] and hydrosilylation,[6] make SAAs an extremely attractive alternative to conventional PGM catalysts.

The performance of alloy catalysts is, however, associated with the structure and morphology of the catalytic surface, and structural modifications specific to the application may be required in order to achieve high catalytic performance. The significance of the dopant and host metal loadings, as well as the way these atoms are distributed on the nanoparticle (NP) facet, have been underlined through computational and experimental studies.[7,8] Dopant atom clustering may be beneficial to the catalytic activity,[8] though the formation of dimers, trimers or islands may result in the deterioration of catalytic performance as a result of the increased susceptibility to CO poisoning.[3] It follows that establishing methods to manipulate the surface structure according to the needs of the specific application is imperative. We recently investigated the structural stability of highly dilute alloys in the presence and absence of CO,[9] demonstrating that the strong binding of CO to dopant metal atoms has a marked influence on the structural stability of highly dilute alloys through effects such as segregation and aggregation.

In this work, we employ Monte Carlo (MC) simulations and Density Functional Theory (DFT) to further examine the most stable structures of PGM alloys with loading up to 5 % in Ag, Au and Cu hosts. We perform DFT calculations on the (111), (100) and corner/edge facets with various dopant atom loadings and CO coverages. We use the energetics determined using DFT to parameterize on lattice MC simulations to elucidate the surface structure of highly dilute NPs of PGM alloys. We establish the most favored facet for dopant atoms, thereby providing a guide for the selection thereof in future theoretical and experimental investigations of catalytic reactions on these alloys. Finally we show how the partial pressure of CO may be manipulated to tune active PGM atom ensemble size and to engineer alloy surface architectures that will be of significant use in catalysis.

References

1. O'Brien, C. P.; Lee, I. C. CO Poisoning and CO Hydrogenation on the Surface of Pd Hydrogen Separation Membranes. *J. Phys. Chem. C* **2017**, *121* (31), 16864–16871.
2. Chung, D. Y.; Kim, H. Il; Chung, Y. H.; Lee, M. J.; Yoo, S. J.; Bokare, A. D.; Choi, W.; Sung, Y. E. Inhibition of CO Poisoning on Pt Catalyst Coupled with the Reduction of Toxic Hexavalent Chromium in a Dual-Functional Fuel Cell. *Sci. Rep.* **2014**, *4* (Iii), 1–5.
3. Liu, J.; Lucci, F. R.; Yang, M.; Lee, S.; Marcinkowski, M. D.; Therrien, A. J.; Williams, C. T.; Sykes, E. C. H.; Flytzani-Stephanopoulos, M. Tackling CO Poisoning with Single-Atom Alloy Catalysts. *J. Am. Chem. Soc.* **2016**, *138* (20), 6396–6399.
4. Pei, G. X.; Liu, X. Y.; Wang, A.; Lee, A. F.; Isaacs, M. A.; Li, L.; Pan, X.; Yang, X.; Wang, X.; Tai, Z.; et al. Ag Alloyed Pd Single-Atom Catalysts for Efficient Selective Hydrogenation of Acetylene to Ethylene in Excess Ethylene. *ACS Catal.* **2015**, *5* (6), 3717–3725.
5. Shan, J.; Lucci, F. R.; Liu, J.; El-Soda, M.; Marcinkowski, M. D.; Allard, L. F.; Sykes, E. C. H.; Flytzani-Stephanopoulos, M. Water Co-Catalyzed Selective Dehydrogenation of Methanol to Formaldehyde and Hydrogen. *Surf. Sci.* **2016**, *650*, 121–129.
6. Miura, H.; Endo, K.; Ogawa, R.; Shishido, T. Supported Palladium-Gold Alloy Catalysts for Efficient and Selective Hydrosilylation under Mild Conditions with Isolated Single Palladium Atoms in Alloy Nanoparticles as the Main Active Site. *ACS Catal.* **2017**, *7* (3), 1543–1553.
7. Liu, P.; Nørskov, J. K. Ligand and Ensemble Effects in Adsorption on Alloy Surfaces. *Phys. Chem. Chem. Phys.* **2001**, *3* (17), 3814–3818.
8. Takehiro, N.; Liu, P.; Bergbreiter, A.; Nørskov, J. K.; Behm, R. J. Hydrogen Adsorption on Bimetallic PdAu(111) Surface Alloys: Minimum Adsorption Ensemble, Ligand and Ensemble Effects, and Ensemble Confinement. *Phys. Chem. Chem. Phys.* **2014**, *16* (43), 23930–23943.
9. Darby, M. T.; Sykes, E. C. H.; Michaelides, A.; Stamatakis, M. Carbon Monoxide Poisoning Resistance and Structural Stability of Single Atom Alloys. *Top. Catal.* **2018**, Accepted, DOI: 10.1007/s11244-017-0882-1.

530452 Unravelling Hydrogenation Barriers for CO_2 Reduction on Nitrogen Doped Zigzag Edges of Graphene

Abstract:

We have used first principles quantum chemistry methods including cluster-continuum solvation modeling and the growing string method to study reaction pathways for CO_2 reduction on cluster models for nitrogen doped zigzag edges of graphene. Our motivation was to better understand the viability of fundamental pathways classified as sequential proton and electron transfers, hydrogen atom transfers, and proton-coupled hydride transfers that would convert CO_2 into CO, formate, and other more reduced products. We analyzed the degree that local solvation environments (i.e. explicit solvent molecules and the proximity of electrolyte ions) would impact the relative energetics of these pathways. This work provides an atomic scale elucidation of energetically viable hydrogenation pathways for the design of renewable energy catalysts.

532015 Overcoming Ammonia Synthesis Scaling Relations with Plasma-Enabled Catalysis

Abstract:

Ammonia synthesis from its elements is only feasible on metal catalysts at elevated temperatures and pressures. These temperature and pressure requirements are known to arise from the constraints imposed by linear scaling relations between the nitrogen activation energy and the adsorption energy of N-containing intermediates---it is not possible for catalysts to simultaneously have a low activation energy for nitrogen dissociation and a weak interaction with surface intermediates [1]. In this talk, I will consider the potential to overcome these constraints and enable high ammonia synthesis activity at low temperatures and pressure by coupling metal catalysts to a non-thermal electric discharge plasma.

We postulate that nitrogen excitations in the plasma phase decrease the energy required to dissociate the robust N-N triple bond. We develop a density-functional-theory-based microkinetic model to incorporate this effect, and parameterize the model using N2 vibrational excitations observed in a dielectric barrier discharge plasma. Two key insights emerge from this model. The first is that ammonia synthesis rates in the presence of the plasma are expected to be greatly enhanced over thermal rates for a given bulk temperature and pressure. Second, optimal catalyst materials and active sites in the presence of plasma excitation are not the same as those for thermal catalysis. The model provides guidance for optimizing catalysts for application with plasmas---ammonia synthesis rates observed in a dielectric-barrier-discharge plasma reactor are consistent with predicted enhancements and predicted changes in the optimal metal catalyst [2].

References

[1] Vojvodic, A. & Norskov, J. K. *National Science Review.* 2, **2015**, 140â€"149.

[2] Mehta, P.; Barboun, P.; Herrera, F. A.; Kim, J.; Rumbach, P.; Go, D. B.; Hicks, J. C.; Schneider, W. F. *Nature Catalysis,* **2018,** 1, 269â€"275.

529597 How Do DFT+U and Hybrids Alter Widely Applied Linear Scaling Relations in Heterogeneous Catalysis?

Abstract:

First-principles electronic structure calculations (e.g. with density functional theory or DFT) provide unique insights into computational materials and catalyst design. However, the approximations made in semi-local DFT models suffer from delocalization error that prevents predictive accuracy on the most compelling materials to study computationally, such as transition metal catalysts. Within surface chemistry, energetics of adsorbates and surfaces will be sensitive to the relative degree of delocalization permitted in a coordination geometry. Thus, more strongly coordinating geometries will be stabilized over weakly interacting ones, and adsorption energies will be overestimated. We carry out a systematic study to identify commonalities in how diverse approaches for approximate delocalization error correction (i.e., DFT+U and hybrid functionals) perform on adsorption energies and surface formation energies of catalytically active, correlated transition metal oxide surfaces. We will present periodic-table-dependent, method-dependent, divergent behaviors on representative rutile-type transition metal dioxides, which are widely employed as catalysts for water splitting. We will discuss how these approaches alter linear scaling relations between different intermediates, and thus identifying materials at the top of the â€œvolcanoâ€ maximum in activity. Time permitting, we will also present our observations on how these same delocalization error corrections alter essential properties of the density in correlated solids.

522729 Grand Canonical DFT Investigation of CO_2 Electroreduction on Noble and Transition Metal Surfaces

Abstract:

Recent advances in the implementation of grand canonical electronic density functional theory enable constant potential calculations for states along the reaction coordinate of electrochemical processes (R. Sundararaman *et al. J. Chem. Phys*. **146**, 114107 (2017)). The algorithm consists of variationally optimizing the number of electrons to converge the grand free energy at fixed potential with the net electronic charge neutralized using continuum solvation method. We tested this technique to investigate CO_2 electroreduction to CO on Ag(111), Au(111), Cu(111), Ir(111), Ni(111), Pd(111), Pt(111) and Rh(111) surfaces and compared the results from those obtained via the widely used computational hydrogen electrode method (J. Norskov, *J. Phys. Chem. B* **108**, 17886 (2004)). The free energy profiles predicted from both schemes yielded similar thermodynamic limiting steps with associated overpotential that are not significantly different from each other. On coinage metals, for example, the first reduction step is the thermodynamic limiting step. Calculations of the kinetic barriers and their potential dependencies are currently underway.

526220 Determination of H_2O-Solvated Cationic Fe(III) Coordination Geometry in Fe-SSZ-13 Using Wavefunction Coupled-Cluster Parameterized Hybrid Density Functional Theory

Abstract:

Zeolites with Fe dispersed as cations have been widely reported as effective industrial and environmental catalysts for a wide range of reactions. Identifying the state of cationic Fe centers is important for rationalization of the corresponding catalytic functions that allows strategy development for further material optimization. Among various geometrical or electronic descriptors, Fe coordination geometry is a geometrical fingerprint that allows identification of ligands and adsorbates attached to Fe under a specific condition, which can be inferred from spectroscopic spectra (e.g. UV-Vis, Extended X-Ray Absorption Fine Structure, etc.), but generally require peak assignment or model fitting, so that conclusions greatly depend on the type of spectroscopy used and individual interpretation of the spectra. An alternative is to explicitly simulate Fe coordination from first-principles using the major workhorse for zeolite modeling, density functional theory (DFT). One of the technical challenges, however, is the accuracy of DFT in probing transition metal coordination chemistry. Recent literature suggests hybrid DFT with parameterized Hartree-Fock exact exchange (HF) incorporation is required to qualitatively reproduce transition metal (e.g. Cu(II)) coordination geometry preference predicted by the benchmarking wavefunction method.

In this work, we focus on H2O-solvated monomeric Fe(III) complexes, which have been shown to be the dominant species in zeolites under ambient conditions. We first benchmarked plane-wave supercell GGA and hybrid HSE functionals against local basis set wavefunction coupled-cluster method to determine optimal HF fraction for modeling coordination of H2O-solvated Fe(III) clusters. We observed that both GGA and HSE with insufficient HF lead to qualitatively erroneous conclusions, highlighting the importance of theory benchmarking. Further, we applied the optimized HSE functionals for zeolite systems. We used SSZ-13 as a model zeolite framework as it has only one symmetry-distinct tetrahedral-site (T-site), which avoids the complexity of decoupling T-site preferences for cationic Fe siting. We examined H2O-solvated Fe(III) charged-compensated by either single or paired Al sites in SSZ-13 and observed the trigonal bipyramidal coordination preference for zeolitic Fe(III) clusters, which is different from its octahedral preference in the aqueous phase. We further performed ab initio molecular dynamics on the preferentially-coordinated complexes for simulations of the radial distribution function (RDF) between Fe and the surrounding heavy atoms (O, Si, and Al). The simulated RDFs are qualitatively distinguishable between single and paired Al sites, potentially allow comparison with EXAFS for Fe center identification at the molecular level. Our findings also highlight a caveat in general of DFT accuracy during theoretical interrogation of transition metal cation coordination relevant problems, e.g. predicting preferred coordination geometry, computing energetics with a change of coordination geometry and/or number between intermediates, etc.

535895 Coverage Dependent Adsorption of Phenol on Pt (111): Estimating the Lateral Interactions Exhibited By Bio-Oil Model Compounds Under Hydrodeoxygenation Reaction Conditions

Abstract:

ABSTRACT

Coverage-Dependent Adsorption of Phenol on Pt (111): Estimating the Lateral Interactions Exhibited by Bio-Oil Model Compounds under Hydro-deoxygenation Reaction Conditions

Neeru Chaudhary, Alyssa J.R. Hensley, Yong Wang, and Jean-Sabin McEwen

Biofuels have gained attention as a long-term, environmentally friendly alternative energy source that has the potential to meet the increasing energy demands from a growing world population. Lignocellulosic biomass can be sufficiently converted to liquid fuels with high yields of about 70-80% through fast pyrolysis.[1] However, the composition of biomass-derived oils are quite different as compared to petroleum oils mostly due to the presence of water and oxygen in the form of phenolic, aldehyde, ketone, and carboxylic compounds, which results in undesired properties that hinder its direct utilization as a transportation fuel.[1, 2] Bio-oil can be upgraded to fuels by catalytic hydro-deoxygenation (HDO) under high pressures and moderate temperatures in the presence of hydrogen,[3] but requires the use of expensive metal catalysts such as Pt, Pd, Ru, or Rh.[2] One way of mitigating this economic drawback is to use inexpensive base metals like Fe and Cu with small amounts of a noble metal, which improves the stability and activity of the base metal.[4, 5] Currently, the HDO mechanism is usually studied at low to medium coverages, for which there are no adsorbate-adsorbate interactions. However, such interactions can affect the chemical kinetics to a great extent.[6, 7] Therefore, it is important to understand not only the effect of the substrate on key HDO ad-species, but also the lateral interaction between ad-species in order to get a clear picture of the surface under realistic reaction conditions.

In this work, we use density functional theory (DFT) to model the coverage dependent adsorption energetics of phenol on Pt (111). Several adsorption sites were tested that varied both the ring position and the functional group position over the surface, as shown in **Figure 1a**. Over the entire coverage range tested, the bridge 30° ring position followed by HCP 0° ring position were the most favorable, with the most favorable functional group position being above surface bridge sites. In general, the change in the ring position has a much larger energetic effect than the functional group position.

Our results show that the phenol adsorption energy varies fairly linearly with the coverage for all the adsorption configurations (**Figure 1b**); however, near saturation the adsorption of phenol slightly deviates from its linear behavior. This could be the result of stronger through-space phenol-phenol repulsions at higher coverages as compared to low coverages where such interactions are almost negligible due to the high separation between ad-species. These through-space repulsions are manifested through a decrease in the dihedral CO bond angle, which decreases from 111° to 107° as the nearest distance between the adsorbates decreases from 8.8 Å to 1.9 Å, From these adsorption energies, we then compute the average differential heat of adsorption, which agrees well with the corresponding experimental data, as shown in **Figure 1c**.[8] This comparison shows that we have accurately captured the coverage dependent adsorption energetics of phenol on Pt (111) within a mean-field model. Since a mean field model can also applied for the adsorption of benzene on Pt(111),[9] this also indicates that such a mean field model can be extended to other aromatic molecules of interest. As such, it provides a simple method for including coverage effects into micro-kinetic models, thereby making such models significantly more realistic. Overall, this work will have a large influence on the current state-of-the-art in modeling in HDO surface reactions by allowing the use of simple models to incorporate coverage effects for large, aromatic compounds.

Figure 1. Adsorption of phenol on Pt (111). **(a)** Most favorable adsorption sites. **(b)** Adsorption energy of phenol at each site as a function of coverage. **(c)** Comparison between the differential heat of adsorption as obtained in a mean field model and the experimental data.

REFERENCES

-

1. Czernik, S. and A.V. Bridgwater, *Overview of Applications of Biomass Fast Pyrolysis Oil.* Energy & Fuels, 2004. **18**(2): p. 590-598.

2. Si, Z., X. Zhang, C. Wang, L. Ma, and R. Dong, *An overview on catalytic hydrodeoxygenation of pyrolysis oil and its model compounds.* Catalysts, 2017. **7**(6): p. 169.

3. Baker, E. and D. Elliott, *Catalytic upgrading of biomass pyrolysis oils*, in *Research in thermochemical biomass conversion*1988, Springer. p. 883-895.

4. Olcese, R., M. Bettahar, D. Petitjean, B. Malaman, F. Giovanella, and A. Dufour, *Gas-phase hydrodeoxygenation of guaiacol over Fe/SiO2 catalyst.* Applied Catalysis B: Environmental, 2012. **115**: p. 63-73.

5. Yu, W., K. Xiong, N. Ji, M.D. Porosoff, and J.G. Chen, *Theoretical and experimental studies of the adsorption geometry and reaction pathways of furfural over FeNi bimetallic model surfaces and supported catalysts.* Journal of Catalysis, 2014. **317**: p. 253-262.

6. Sabbe, M.K., G. Canduela-Rodriguez, J.-F. Joly, M.-F. Reyniers, and G.B. Marin, *Ab initio coverage-dependent microkinetic modeling of benzene hydrogenation on Pd (111).* Catalysis Science & Technology, 2017. **7**(22): p. 5267-5283.

7. Park, Y.K., P. Aghalayam, and D.G. Vlachos, *A generalized approach for predicting coverage-dependent reaction parameters of complex surface reactions: application to H2 oxidation over platinum.* The Journal of Physical Chemistry A, 1999. **103**(40): p. 8101-8107.

8. Carey, S.J., W. Zhao, Z. Mao, and C.T. Campbell, *Energetics of Adsorbed Phenol on Ni(111) and Pt(111) by Calorimetry.* The Journal of Physical Chemistry C (submitted), 2018.

9. Wong, B., G. Collinge, A.J.R. Hensley, Y. Wang, and J.-S. McEwen, *Increasing the Modeling Accuracy of the Desorption Kinetics Through the Inclusion of Coverage Effects: Benzene on Pt (111) and Pt3Sn (111) from First Principles.* Progress in Surface Science (submitted), 2018.

538582 Nonlinear Manifold Learning of Nucleosome Dynamics from Molecular Simulation

Abstract:

The identification of important dynamical motions remains a challenge in molecular simulations of complex systems. Here, we use a nonlinear manifold learning technique known as the diffusion map to extract key dynamical modes from molecular simulations of the nucleosome, a complex biomolecular system consisting of 147 base pairs of DNA wrapped around a disc-shaped histone octamer. We demonstrate that diffusion maps are effective at extracting motions consistent with those previously found through a detailed free energy analysis (e.g. DNA translocation along the histone complex). We also show that diffusion maps can reveal more subtle features, including the formation of looping conformations, in which DNA bulges away from the histone complex, as well as motions involved in DNA breathing, where DNA spontaneously unwraps from the histone octamer.

536589 Machine Learning Predicts Functional Classes of Family 7 Glycoside Hydrolases with High Accuracy

Abstract:

Glycoside hydrolases (GH) are a class of enzymes that catalyze the hydrolysis of glycosidic bonds in saccharides. They are utilized in industries, such as the biofuel and textile industries, for enzymatic degradation and reorganization of saccharides. GHs are presently classified into 152 families based on sequence identity. Family 7 glycoside hydrolases (GH7s) are predominantly found in fungi and are often the largest composition by mass of the secretomes of cellulolytic fungi. In the biofuel industry, GH7s are the primary components of the enzymatic cocktails used in cellulose degradation. GH7s fall into one of two classes: cellobiohydrolases (CBHs) or endoglucanases (EGs). GH7 CBHs hydrolyze cellulose processively, i.e. they carry out multiple catalytic steps without dissociating from the substrate. GH7 EGs, on the other hand, are non-processive and dissociate from the substrate after hydrolyzing a glycosidic bond. Processive GH7s (CBHs) have become a focus of research because they provide the greatest hydrolytic potential in enzymatic cellulose degradation. As many of the known GH7 sequences have not yet been classified in terms of activity, we have set out to develop a predictive approach for classifying GH7 activity. We first retrieved a large and diverse set of 1,521 GH7 sequences from the genomic databases. The functional classes (i.e. CBH or EG) are reported for only about 30% of these GH7s. We trained multiple machine learning classifiers (decision tree, SVM, naïve Bayes and logistic regression) using known structural differences between GH7 CBHs and EGs as features. We determined, using Monte Carlo cross validation, that the overall accuracy of the machine learning classifiers ranges from 95 to 97%, suggesting that GH7 functional class can be readily predicted from sequence alone.

534508 Implementation and Automation of a Hierarchical Graph Based Approach for Extracting Coarse-Grain Mapping Operators

Abstract:

Coarse-grain (CG) molecular dynamics (MD) simulation allows us to overcome the limitations of all-atom (AA) MD simulation in terms of length and time scales. While CG-MD has been successfully applied to various systems like protein folding, there is still a need for a general guideline to determine the choice of CG mapping operators. Recently we have proposed a technique of incorporating multiple CG representations of a molecule within a single hierarchical graph. With the ability to extract valid CG mapping operators from the graph and relevant thermodynamic property like entropy driving the selection of one among the various mapping operators extractable from the hierarchical graph, this technique is a step towards automating the selection of CG mapping operators. As a proof of concept we have built a hierarchical graph for methanol which encodes all symmetry preserving CG mapping operators. We also demonstrated the feasibility of using the graph for automated CG mapping operator selection by using the uniform entropy flattening method to select CG mapping operators for methanol. In this talk, we will describe the implementation of the algorithm along with an automated pipeline framework to create a database of benchmark CG simulations. The discussion will include different thermodynamic measures to compare the CG and the AA performances. The large data set of CG simulations will aid in the development of standardized and scalable CG mapping strategy that is system independent.

521801 Machine Learning Algorithms for High-Throughtput Chemical Sensing Using Liquid-Crystals

Abstract:

Security, health, and environmental monitoring requires the development of chemical sensing technologies that can be used in-situ and with limited equipment and human intervention. The impact of such technologies when deployed at a large scale can be significant; for instance, the U.S. Department of Energy Savannah River laboratory analyzes over 40,000 groundwater samples per year at a cost of $1,000 per sample ($40 million per year) [1]. Liquid crystals (LCs) are fluid phases with preferred molecular orientations, a so-called director, that undergo surface-driven ordering transitions in the presence of chemical species such as organophosphonates [2, 3, 4, 5, 6], chlorine, ammonia, and hydrogen sulfide [7]. The optical characteristics (features) of the LC transitions can be tailored and exploited to design chemical sensors. For instance, LCs can be designed to assume homeotropic (perpendicular) orientations on surfaces decorated with metal salts [2, 4, 8]. Chemical species that diffuse into the LCs and bind more strongly to the metal cations than the LC functional groups will trigger a transition of the LCs orientation from homeotropic to planar (see Figure 1) [9, 10, 11]. It is possible to manipulate the selectivity and response characteristics (e.g., dynamics) of the LC by tuning the binding energies of the LC functional groups (e.g., nitrile and pyridine groups) to the surface (e.g., Fe+3, La3+) [9, 10]. For instance, ordering transitions of LC sensors fabricated using a nematic LC called 4-cyano- 4'-penthylbiphenil (5CB) and surfaces presenting aluminum perchlorate salts have been studied in [12, 13, 14].

In this work, we present a machine learning (ML) framework to optimize the specificity and speed of LC-based chemical sensors. Specifically, we demonstrate that ML techniques can uncover valuable feature information from surface-driven LC orientational transitions triggered by the presence of different gas-phase analytes (and the corresponding optical responses) and can exploit such feature information to train accurate and automatic classifiers. We demonstrate the utility of the framework by designing an experimental LC system that exhibits similar optical responses to a stream of nitrogen containing either 10 ppmv dimethyl-methylphosphonate (DMMP) or 30% relative humidity (RH). The ML framework is used to process and classify thousands of images (optical micrographs) collected during the LC responses and we show that classification (sensing) accuracies of over 99% can be achieved. For the same experimental system, we demonstrate that traditional feature information used in characterizing LC responses (such as average brightness) can only achieve sensing accuracies of 60%. We also find that high accuracies can be achieved by using time snapshots collected early in the LC response, thus providing the ability to create fast sensors. We also show that the ML framework can be used to systematically analyze the quality of information embedded in LC responses and to filter out noise that arises from imperfect LC designs and from sample variations. We evaluate a range of classifiers and feature extraction methods and conclude that linear support vector machines are preferred and that high accuracies can only be achieved by simultaneously exploiting multiple sources of feature information.

References

[1] Clifford K Ho, Alex Robinson, David R Miller, and Mary J Davis. Overview of sensors and needs for environmental monitoring. Sensors, 5(1):4–37, 2005.

[2] K-L Yang, Katie Cadwell, and Nicholas L Abbott. Contact printing of metal ions onto carboxylate-terminated self-assembled monolayers. Advanced Materials, 15(21):1819–1823, 2003.

[3] Kun-Lin Yang, Katie Cadwell, and Nicholas L Abbott. Mechanistic study of the anchoring behavior of liquid crystals supported on metal salts and their orientational responses to dimethyl methylphosphonate. the Journal of Physical Chemistry B, 108(52):20180–20186, 2004.

[4] Katie D Cadwell, Mahriah E Alf, and Nicholas L Abbott. Infrared spectroscopy of competitive interactions between liquid crystals, metal salts, and dimethyl methylphosphonate at surfaces. The Journal of Physical Chemistry B, 110(51):26081–26088, 2006.

[5] Maricar L Bungabong, Peng Bin Ong, and Kun-Lin Yang. Using copper perchlorate doped liquid crystals for the detection of organophosphonate vapor. Sensors and Actuators B: Chemical, 148(2):420–426, 2010.

[6] Santanu Kumar Pal, Claribel Acevedo-Velez, Jacob T Hunter, and Nicholas L Abbott. Effects of divalent ligand interactions on surface-induced ordering of liquid crystals. Chemistry of Materials, 22(19):5474–5482, 2010.

[7] Sheila E Robinson, Bart A Grinwald, Laura L Bremer, Kurt A Kupcho, Bharat R Acharya, and Patrick D Owens. A liquid crystal-based passive badge for personal monitoring of exposure to hydrogen sulfide. Journal of occupational and environmental hygiene, 11(11):741–750, 2014.

[8] Jacob T Hunter and Nicholas L Abbott. Adsorbate-induced anchoring transitions of liquid crystals on surfaces presenting metal salts with mixed anions. ACS applied materials & interfaces, 6(4):2362–2369, 2014.

[9] Tibor Szilv´asi, Luke T Roling, Huaizhe Yu, Prabin Rai, Sangwook Choi, Robert J Twieg, Manos Mavrikakis, and Nicholas L Abbott. Design of chemoresponsive liquid crystals through integration of computational chemistry and experimental studies. Chemistry of Materials, 29(8):3563– 3571, 2017.

[10] Huaizhe Yu, Tibor Szilv´asi, Prabin Rai, Robert J Twieg, Manos Mavrikakis, and Nicholas L Abbott. Computational chemistry-guided design of selective chemoresponsive liquid crystals using pyridine and pyrimidine functional groups. Advanced Functional Materials, 2018.

[11] Luke T Roling, Jessica Scaranto, Jeffrey A Herron, Huaizhe Yu, Sangwook Choi, Nicholas L Abbott, and Manos Mavrikakis. Towards first-principles molecular design of liquid crystalbased chemoresponsive systems. Nature communications, 7:13338, 2016.

[12] Daniel S Miller, Rebecca J Carlton, Peter C Mushenheim, and Nicholas L Abbott. Introduction to optical methods for characterizing liquid crystals at interfaces. Langmuir, 29(10):3154–3169, 2013.

[13] Alireza Hassanzadeh and Robert G Lindquist. Liquid crystal sensor microchip. IEEE Sensors Journal, 12(5):1536–1544, 2012.

[14] Rahul R Shah and Nicholas L Abbott. Orientational transitions of liquid crystals driven by binding of organoamines to carboxylic acids presented at surfaces with nanometer-scale topography. Langmuir, 19(2):275–284, 2003.

533502 Materials Informatics for Process Optimization: Case Studies Using P3HT and PP Composites

Abstract:

The traditional paradigm in materials process optimization is energy-intensive and time consuming. Generally, single factor, trial-and-error based experimentation is used to screen a wide array of complex processing conditions. This generates a vastly multidimensional design space that is impossible to fully characterize. Furthermore, results are difficult to reproduce because of a lack of standardization due to the exploratory nature of the field. As a result, it is challenging to develop global, robust process-structure-property relationships to generate novel material formulations. Instead, informatics based methodologies can leverage existing knowledge from the literature to guide high-throughput screening of promising candidate experimental conditions. In this work two case studies of applying materials informatics methodologies to polymer systems will be presented.

In the first study, the impact of processing conditions on the organic field effect mobility of the semiconducting polymer, poly(3-hexylthiophene) (P3HT), will be presented. A database of over 200 devices was created to identify significant trends and to identify an optimal operating space. Processing conditions were sorted and filtered to identify a standard device, with reported mobility values that span two orders of magnitude. Furthermore, both integer and categorical processing conditions resulting in mobility values exceeding 0.1 $cm^2/V\text{-}s$ were identified to reduce the dimensionality of the design space. This enabled the generation of new hypotheses for future high-throughput experimentation.

In the second study, a database with over 140 entries quantifying the development of polypropylene and talc composites for high strength materials was created. However, this database suffered from inconsistent reporting of processing conditions. Physical based models, for example relating melt flow index to molecular weight, were proposed to help fill in missing data. Varying thresholds of Youngâ€™s Modulus were applied to identify the span of processing conditions that result in high strength composites. In both studies quantified analysis of all processing conditions has identified key design spaces to enable high through-put experimentation.

537478 Learning Many-Body Molecular Interactions from Machine Learning

Abstract:

The accurate representation of multidimensional potential energy surfaces is a necessary ingredient for realistic computer simulations of molecular systems. It has recently been shown that chemical and spectroscopic accuracy can be achieved with analytical potential energy functions (PEFs) rigorously derived from many-body expansions. In this contribution, we demonstrate the equivalence between permutationally invariant polynomials, neural networks, and Gaussian approximation potentials in representing lower-order many-body terms of molecular interactions in aqueous systems from the gas to the condensed phase. Besides demonstrating the synergy between high-quality electronic structure data and machine-learning techniques for developing transferable PEFs, our results show that machine learning provides a powerful tool for learning the underlying physics of many-body molecular interactions.

531504 Combining Machine Learning and Evolutionary Computing for Accelerating Materials Design

Abstract:

Machine learning models have been progressively adopted by the materials science community to accelerate design of materials with targeted properties. However, in the search for new materials exhibiting properties and performance beyond that previously achieved, machine learning approaches are frequently limited by two major shortcomings. First, they are intrinsically interpolative. They are therefore better suited to the optimization of properties within the known range of accessible behavior than to the discovery of new materials with extremal behavior. Second, they require the availability of large datasets, which in some fields are not available and would be prohibitively expensive to produce. Here we describe a new strategy for combining genetic algorithms, neural networks and other machine learning tools, and molecular simulation to discover materials with extremal properties in the absence of pre-existing data. Predictions from progressively constructed machine learning tools are employed to bias the evolution of a genetic algorithm, with fitness evaluations performed via direct molecular dynamics simulation. We survey several materials design problems we have addressed with this framework. Specific examples for designing soft materials and 2D organic materials will be presented.

535265 Predicting Colloidal Crystals from Shapes Via Inverse Design and Machine Learning

Abstract:

A fundamental challenge in materials design is linking building block attributes to crystal structure. Addressing this challenge is particularly difficult for systems that exhibit emergent order, such as entropy-stabilized colloidal crystals. We combine recently developed techniques in inverse design with machine learning to construct a model that correctly classifies the crystals of more than ten thousand polyhedral shapes into 13 different structures with a predictive accuracy of 96% using only two geometric shape measures. With three measures, 98% accuracy is achieved. We test our model on previously reported colloidal crystal structures for 71 symmetric polyhedra and obtain 92% accuracy. Our findings (1) demonstrate that entropic colloidal crystals are controlled by surprisingly few parameters, (2) provide a quantitative model to predict these crystals solely from the geometry of their building blocks, and (3) suggest a prediction paradigm that easily generalizes to other self-assembled materials.

522234 Autonomous Crystal Structure Characterization with Neighborhood Graph Analysis

Abstract:

We have previously developed a template-free method for characterizing local crystalline structure called Neighborhood Graph Analysis. Unlike conventional methods, our technique uses nonlinear manifold learning to infer structural relationships directly from particle tracking data. While it provides exceptional flexibility and discriminating power, in the past our graph-based classifier was prohibitively slow to evaluate, making it impractical for everyday use. We present a modification to the original method which results in nearly identical results while speeding up the calculation by three orders of magnitude. As with the original technique, we are able to characterize particles near defects, grain boundaries, and interfaces of close-packed crystal structures. However, the increased speed of our new algorithm allows us to extend the analysis to more exotic structures such as ices, clathrates, perovskites, and pyrochlores. We can even handle superlattices composed of two or more species. We demonstrate how the resulting analysis can be used to enhance structural analysis whether the relevant structures are known or not.

537653 The Physical Analytics Pipeline - a Bayesian Optimization of the Hybrid Organic-Inorganic Perovskite Compositional Space

Abstract:

Hybrid Organic-Inorganic Perovskites (HOIPs) are an exciting class of photovoltaic materials due to their combination of high solar cell efficiency and the ability to be fabricated by a low-energy room temperature solution process. Perovskites are classified by their characteristic ABX3 structure, where A is an organic (or inorganic) cation (e.g., methyammonium, formamadinium or Cs), B is a metal cation (typically Pb or Sn), and X is a halide (where X= I, Br, Cl). Choice of solvent is also very important in the fabrication process and can be one of perhaps eight common solvent choices. With all these possible choices of ABX3, including mixing metal cations, mixing halides and use of binary solvent blends, the HOIP compositional space quickly becomes unwieldy, with at least 500,000 possible combinations. Experimental exploration of this space is Edisonian and limited by time and resources and a lack of underlying rational design guidelines. Even computational approaches are limited by the lack of available force fields for the multi-species HOIP systems, which restricts the calculations to expensive quantum mechanical ones. The Physical Analytics pipeLine (PAL), which we have developed, tackles this problem by bridging Gaussian Process Bayesian Optimization with computational methods to minimize the number of necessary calculations needed to locate the global maximum. As a proof of concept, PAL has been implemented for pure and mixed halide perovskites, and is benchmarked against the Sequential Model-based Algorithm Configuration (SMAC). PAL performs extremely well in comparison to SMAC, and is robust upon replication with different initial training sets. Further improvements to the PAL model have been made to take into account the ability to combine a variety of information sources, with different accuracy and cost, allowing us to use cheaper, less accurate calculations to aid in improving our predictions, whilst minimizing the total cost of optimization. This combination of exploration and exploitation allows for efficient location of broad parameter spaces where data may be sparse.

542620 The Impact of Shale Gas and Oil on the Chemical Industry

Abstract:

Lower alkanes and aromatics are very useful as both fuels and chemical raw materials. Occasionally supply disruptions constrained availability but also encouraged investigation of alternative feedstocks and chemistries. Recent new technologies for the production of unconventional gas and oil sources from tight shale formations significantly increased availability of both methane and condensate, radically altered process economics, resulted in the construction of much new intermediates capacity, and multiplied the ways traditional and alternative chemistries might be exploited. This presentation will outline historically how natural gas components became useful as chemical feedstocks, how government regulations occasionally led to perturbations in availability and price, how supply crises led to the development of alternative chemistries, catalysts, and ultimately the exploitation of shale resources. More specifically, will be discussed how excess coproducts, especially condensate, from increasing shale gas and oil production for energy are having significant unexpected impacts on the chemical process industries and opportunities for new process chemistry research.

541186 Sustainable Energy and Chemicals: Past, Present, and Future

Abstract:

The last 40 years has seen the U.S. energy situation vacillate from gasoline rationing and concerns over Peak Oil, to projected energy and feedstock prices that were noncompetitive in global markets for chemicals production, to the economic revolution of frack gas and light tight oil. The importance of low cost energy and feedstocks in driving economic opportunity will be discussed relative to this history, along with the importance of sustainability in decision making to gain stakeholder buy-in and acceptance. Strategies for managing career paths and competencies despite changing targets and high uncertainty will be addressed.

542618 Disruptions: What the Future May Hold

Abstract:

Disruptive forces are all around us. They change the way we communicate and work. They change the way we develop goods and services. They also change the types of products that are manufactured. Examples such as the way in which we produce and consume energy or the feedstocks we use to produce product have seen many changes over the course of the last few decades, and these disruptions are expected to continue. The chemical engineering discipline and the expectations for a chemical engineerâ€™s skills has also changed and adapted as a result of these disruptive forces. A discussion of the broad themes associated with these forces and the implications that they may have to chemical engineers in the future will be discussed.

542621 Geopolitical Factors Influencing the Evolution of the Chemical Industry

Abstract:

The basic chemicals and plastics industry is built on a simple principle: find a cheap feedstock and convert it into something more valuable. However, the right feedstock depends on place and time. Certain policies and actions of various governments have influenced the evolution and profitability of the chemical industry during the past 50 years, sometimes in surprising ways. For example, high oil prices and US government funding encouraged the development and commercialization of coal gasification technology during the late 1970s and early 1980s. This enabled China to develop a competitive basic chemical industry using coal as a feedstock during the last 25 years. Similarly, an abrupt increase in oil prices during the first decade of the 21st century encouraged the present shale gas revolution, causing a major shift in the economy of producing basic chemicals and polymers in the US. If history is our guide, then what governments choose to fund and regulate may be as important as market forces in determining the shape of our industry and profession in the future.

542622 Agility & Resilience: How to Maintain Career Competitiveness in the Changing Chemical Industry

Abstract:

The world around is changing. The shift to protectionism, global instability and price volatility of raw materials and feedstocks has the potential to wreak havoc on the chemical industry. With the revolution of shale gas in North America and the sudden drop in feedstock costs, the chemical sector in the US has reasserted its historic dominance. Learn how the role of a chemical engineer can contribute to industry growth and resilience, while adapting to allow for personal agility throughout the career life cycle.

537824 Development of Interface Potential Based Methods for Calculating the Wetting Properties of Complex Systems

Abstract:

The behavior of fluids at solid surfaces plays a significant role within numerous natural phenomena and industrial applications. To this end, understanding and predicting the interfacial properties of these systems is crucial for the development of emerging technologies. Within our group, we have developed an interface potential approach for determining interfacial properties. In previous work, we performed simulations within a grand canonical ensemble and calculated the interface potential using transition matrix Monte Carlo methods. However, there are limitations to working within this ensemble, particularly when applying the technique to complex molecules and working at relatively low temperatures. Recently, we developed a means for calculating the interface potential using an isothermal-isobaric ensemble (NPT) using molecular dynamics. In this presentation, we discuss how to implement this approach within a molecular dynamics framework. We perform simulations in the canonical ensemble at different volumes to sample states. We use the forces acting on the substrate to construct an interface potential. Expanded ensemble techniques are used to evaluate interfacial properties over a range of temperatures and surface characteristics. Molecular dynamics simulations are performed using the LAMMPS package. Results are presented for water on a silica surface and 1,3-diMethylImidazolium tetraFluoroBorate ([C1MIm]BF4) on a graphite surface. The results obtained using molecular dynamics are compared with those obtained from grand canonical Monte Carlo simulation.

522953 Molecular Dynamics Analysis of Salt Effect on Anti-Agglomerant Surface Adsorption in Natural Gas Hydrates

Abstract:

The effect of salt on the performance of anti-agglomerant (AA) molecules is AA-dependent, and in most cases salinity improves the performance of AAâ€™s. There is no consensus on saltâ€™s effect on the AA-hydrate interaction, but understanding the influence of salt on AA-hydrate interaction mechanisms and energetics could be a step towards designing better AAâ€™s with reduced experimental trial-and-error in the innovation process. We have used molecular dynamics simulations to examine the surface adsorption of a model anti-agglomerant inhibitor molecule binding to a sII methane-propane hydrate in environments of different salinity. From our simulation data, we were able to identify the preferred binding sites on hydrate and characterize the equilibrium binding configurations. In addition, for a subset of these binding configurations, we calculated the standard binding free energy in different concentrations of brine using the potential of mean force (PMF) based free energy calculations. We demonstrate that in higher salinity environments, the surface adsorption of the anti-agglomerants is enhanced through two distinct mechanisms. First, the salt decreases the solubility of the anti-agglomerant in the solution, which increases the thermodynamic driving force for surface adsorption. Second, the salt ions create a negatively charged interfacial layer close to the hydrate surface that effectively solvates the cationic head of the anti-agglomerant molecule, creating an energetically favorable bound state for the anti-agglomerant molecule. Quantitatively, we find that the presence of 3.5 wt% and 10 wt% NaCl decreases the standard binding free energy of the long hydrocarbon tail binding configuration by 0.8 and 1.4 kcal/mol, and decreases the standard binding free energy of the cationic head binding configuration by 1.5 and 3.3 kcal/mol, respectively. Finally, it decreases the standard binding free energy of simultaneous head and tail binding configuration by 1.9 and 4.3 kcal/mol, respectively.

529854 Adsorption and Self-Assembly of Surfactant Molecules on Metallic Surfaces Studied Using Molecular Simulations

Abstract:

Surfactant molecules are known to adsorb and self-assemble on to metal-water interfaces. While adsorbed surfactant films are employed for many technological applications, such as corrosion inhibition and electrochemical blocking of electrodes, the molecular nature of this phenomenon is only poorly understood. From our coarse-grained simulations, we reveal that the traditional depiction of surfactant adsorption is imprecise on many fronts. We show that lateral interactions between alkyl tails of surfactant molecules play an important role in adsorption and self-assembly; the nature of the adsorbed morphology of surfactant films depends on molecular geometry; the adsorbed surfactant films are patchy with a significant fraction of the polar heads pointing towards the solution; and the magnitude of the free energy barrier to formation of self-assembled monolayers depends on the strength of polar group â€" surface affinity. These results, along with their implications in optimal design of surfactant molecules for different applications, will be discussed in this talk.

533599 Rationalizing Stability and Doping of Atomically Precise Ligand-Protected Metal Nanoclusters

Abstract:

Ligand-protected colloidal nanoclusters of precise size, shape, and composition have attracted interest as model systems in applications from medical imaging to catalysis. Though several thiolate-protected metal clusters have been determined with atomic precision, relatively little is understood of what underlying physics result in their remarkable stability and resultant colloidal monodispersity. Herein we highlight our work [1-4] proposing a new thermodynamic stability theory based on cluster interfacial energetics. Additionally, we extend this theory to rationalize heterometal doping energetics, concentrations, and preferred dopant sites within several bimetallic gold-based nanoclusters. Beyond this new theory, we tie experimental doping observations to the energetics of reactions between nanoclusters and solution-phase precursor complexes. Overall, we present work that further develops understanding of stability and doping atomically-precise metal nanoclusters that can be used for predictions of synthesizable nanoclusters for targeted applications.

[1] M. G. Taylor, G. Mpourmpakis, Thermodynamic Stability of Ligand-Protected Metal Nanoclusters. **Nature Communications** 8, 15988 (2017).

[2] Q. Li et al., Reconstructing the Surface of Gold Nanoclusters by Cadmium Doping. **Journal of the American Chemical Society** 139, 17779-17782 (2017).

[3] Q. Li et al., Molecular â€œsurgeryâ€ on a 23-gold-atom nanoparticle. **Science Advances** 21, 1-8 (2017).

[4] Q. Li et al., Site-selective substitution of gold atoms in the $Au_{24}(SR)_{20}$ nanocluster by silver**. Journal of Colloid and Interface Science** 505, 1202-1207 (2017).

531192 Influence of Chlorine Adsorption on the Thermodynamic Wulff Shape of Ag Nanocrystals

Abstract:

It has been shown experimentally that chlorine can promote the formation of monodisperse Ag nanocubes with sharp corners and edges in the polyol synthesis. However, the role of Cl in the shape-control mechanism of Ag nanocubes remains elusive. Here, we examine how the adsorption of Cl influences the thermodynamic Wulff shape of Ag nanocrystals, where we use density functional theory and ab initio thermodynamics to calculate the surface energy as a function of the Cl chemical potential / solution-phase concentration of Cl. At each Cl chemical potential, the minimum surface energies of Cl-adsorbed Ag(100) and Ag(111) surfaces at different Cl coverages are used to construct the Wulff shape of Ag nanocrystals. Our calculations show that Cl adsorption can change the ratio of surface energies such that the formation of cubes with small truncated corners is thermodynamically favorable, in contrast with the Wulff shape of bare Ag surfaces which is a cubooctahedron. In addition, we find that mixed substitutional+on-surface adsorption becomes more stable than on-surface adsorption at high Cl coverages, which could facilitate the formation of AgCl structures at large solution-phase concentrations of Cl. Our study indicates that the adsorption of Cl can impart significant thermodynamic-shape change, favoring the formation of Ag nanocubes.

539717 Computational Studies of Novel Structure-Directing Agents and Crystal Growth Modifiers for Zeolite Catalysts

Abstract:

Computational Studies of Novel Structure-Directing Agents and Crystal Growth Modifiers for Zeolite Catalysts

1. John Clark and Jeremy C. Palmer

Department of Chemical and Biomolecular Engineering, University of Houston, Houston, Texas, 77204, USA

ABSTRACT

Designing nanoporous zeolite materials with tailored properties is critical to industrial applications ranging from fluid separations and ion exchange to hydrocarbon cracking and methanol reformation. Organic structure-directing agents (OSDAs) are commonly used to facilitate the formation of specific zeolite framework types. Effective OSDAs exhibit a size and shape that is commensurate with the geometry of porous channels and cages of zeolites, thereby directing their formation *in situ*[1]. Zeolite growth modifiers (ZGMs) can also be used in conjunction with OSDAs to further tune material properties. Whereas OSDAs play a critical role in zeolite polymorph selection, ZGMs are molecular additives that influence bulk crystal shape and size by adsorbing to specific crystal faces and altering anisotropic rate(s) of growth during synthesis[1]. Although the combination of OSDAs and ZGMs allows for unprecedented control over zeolite properties to be achieved, identifying new and inexpensive compounds suitable for industrial scale synthesis remains a major challenge.

We present results from three recent combined computational and experimental studies of novel OSDAs and ZGMs for designing improved zeolite catalysts. In the first study[2], we show that a common surfactant (cetrimonium bromide) can be used as an alternative to the standard OSDA (tetrapropylammonium) for stabilizing ZSM-5, which is a widely used industrial catalysts. In the second study, we demonstrate that an unusual synergistic effect between a pair of OSDAs leads to the formation of mordenite nanocrystals, which cannot be produced using either OSDA alone. Finally, we examine the effect of several alcohol-based ZGMs on the shape and size of TON and LTL structured zeolite crystals. In each case, advanced molecular simulation techniques are used to elucidate the mechanisms by which the specific OSDAs and ZGMs influence zeolite stability and growth. Predictions from the simulations are shown to be in agreement with zeolite synthesis, NMR, and immersion calorimetry experiments.

References

[1] A. I. Lupulescu, M. Kumar and J. D. Rimer, "A Facile Strategy to Design Zeolite L Crystals with Tunable Morphology and Surface Architecture", J. Am. Chem. Soc., **2013**, 135, 6608-6617.

[2] A. Chawla, R. Li, R. Jain, R. J. Clark, J. G. Sutjianto, J. C. Palmer and J. D. Rimer, "Cooperative Effects of Inorganic and Organic Structure-Directing Agents in ZSM-5 Crystallization", Mol. Syst. Des. Eng., **2018**, 3, 159-170.

522677 Gas Transport at the Polymer-Zeolite Interface Using Atomistic Simulations

Abstract:

Membrane based separation is considered as efficient, productive, and environmentally friendly processes that operate in a continuous fashion. The development of mixed matrix membranes (MMM) for gas separation is an active and rapidly growing area of research, due to the challenges with the current spectrum of polymeric and inorganic membranes. MMMs have been conventionally prepared by incorporating inorganic fillers such as zeolites, metal organic framework (MOFs) and carbon nanotubes (CNTs) in a continuous polymer matrix. However, the ultimate success of these advanced membranes depends on the material selection and interface defect elimination. Hence, nanoscale understanding of polymer structure near a surface and gas transport at the interface is critical to the design of these advanced gas separation technologies. Here we report a comparison of the adsorption and transport characteristics of CO_2 and CH_4 as single components in polyimide (PI), MFI zeolite and PI-MFI zeolite composite membrane as a function of temperature in the range of 300-500 K using equilibrium Molecular Dynamics (EMD) simulations. It is seen that incorporation of MFI zeolite into PI results in the formation of densified polymer layers near the surface, having thickness around 1.2 nm, before bulk-like behavior of the polymer is attained. This region offers an extra resistance to the gas diffusion especially for the gas with larger kinetic diameter, CH_4 in this case, thus improving the CO_2/CH_4 kinetic selectivity in the PI-MFI composite membrane. Further, we find that the kinetic selectivity of CO_2 over CH_4 in the rigidified region increases with temperature. Furthermore, the temperature dependence of the collective diffusivity of CO_2 and CH_4 follows Arrhenius behavior in PI, MFI zeolite and PI+MFI hybrid membranes. Also, it is seen that crystal size has little effect on the polymer structure and gas transport at the polymer-filler interface. In addition, to extract the isotherms for gas adsorption, we implemented a two-step methodology by performing Grand Canonical Monte Carlo (GCMC) simulations coupled with EMD simulations by considering the dynamics and structural transitions in the polymer matrix upon gas adsorption, and investigated the resulting adsorption isotherms of CO_2, CH_4 in pure and composite polymer membranes. The sorption results indicate that incorporation of MFI zeolite into PI improves the adsorption selectivity of CO_2 over CH_4. A significant increase in CO_2/CH_4 selectivity as well as the gas permeability is observed in the PI-MFI composite membrane compared to the pure PI polymer membrane, which is correlated with the high selectivity of the rigidified interfacial layer in the polymer.

rigidified interface

gas transport

MFI surface

PI polymer

532441 Comparative Analysis of Surface Configurations of CO Adsorbed on hcp and fcc Cobalt for the Fischer-Tropsch Synthesis

Abstract:

The first step in Fischer-Tropsch (FT) synthesis over Co catalysts is carbon monoxide (CO) adsorption, which by and large out-competes dissociative hydrogen adsorption. Along with the fact that CO dissociation is not easy on the basal planes of Co catalysts, this fact implies that associatively adsorbed CO will likely have time to equilibrate with the imposed CO atmosphere and take on configurations relevant to the subsequent steps of the reaction. We present here a density functional theory (DFT) based investigation of these relevant configurations, taking into account the lateral interactions of CO through the construction of highly-predictive lattice gas models. Since both the face centered cubic (fcc) and hexagonal close packed (hcp) phases of Co have been shown to exhibit different FT activity, we compare these LG models and configurations as a function of the fcc and hcp phases; the primary difference being a large change in site-to-site distance (i.e. different lattice strains).

By utilizing the Alloy Theoretic Automated Toolkit to generate a large variety of extended Co(111) and Co(0001) surfaces with different coverages and configurations of CO, we construct a large library of CO/Co(111) and CO/Co(0001) structures. A LG model is fit to these data and subsequently optimized to produce the most predictive model possible for each system. The results reveal a non-trivial change in the first nearest neighbor (NN) interaction energy, reflected in a much more favorable full-coverage in the hcp surface (see Figure 1). However, we are able to show that the ground states of these two systems are largely unchanged despite this difference. Our results facilitate experimentally relevant comparisons of the two systems and indicate what kind of surface structures one should expect to be present in the early stages of FT synthesis given an hcp vs. fcc Co-based catalyst.

Figure 1. Surface energies and associated LG predicted energies for the hcp and fcc Co systems.

526116 Atomistic Modeling Strategies for Solid Electrolyte Interphase Formation and Properties in Lithium Ion Batteries

Abstract:

Lithium ion batteries have driven a technological revolution in consumer electronics and power tools, while also inspiring the development of fully electric vehicles. However, there remains a divide between their theoretical and practical performance as gains in capacity, lifetime, and charge rate are demanded for the evolution of the technologies they power. Many of the practical hurdles lie at the interfaces between the electrodes and the liquid electrolyte, in particular, the dynamic nature of the anode-electrolyte interface plays a key role in the determination of battery performance. The *in situ* formation of an ionic conducting and electronically insulating layer known as the solid electrolyte interphase (SEI) on the anode is critical to robust battery performance, yet a deep fundamental understanding of the process by which it is formed remains elusive. Due to the size and time scales of the electrolyte-electrode interfacial phenomena, limited experimental approaches may provide the appropriate resolution. Computational methods, however, may be applied at spatial and temporal scales often unavailable in experiments. This work details a computational approach to studying the formation of the SEI using atomistic simulations at explicit electrode-electrolyte interphases. Through a multi-scale approach, we have explored the formation of the SEI at various stages including: the reorganization of the electrolyte at a charged graphite electrode surface prior to the first reduction reaction, the transport behavior of the reaction intermediates near the electrode prior to SEI formation, the accumulation of reaction products/intermediates, and the formation of a solid layer on the graphite surface. At each stage of the process, interplay between the electrode-electrolyte and electrolyte-electrolyte interactions dictates behaviors at the molecular level which determine the structure, and ultimately the properties of the SEI, and therefore the LIB. In this talk, we will present some of our recent findings including what we believe to be key descriptors in to predicting these behaviors at the various stages of SEI formation. This improved understanding will hopefully provide design principles for future investigation into forming a stable SEI.

535024 Geometric Surfactancy Probed By Molecular Dynamics Simulation of Lennard-Jones Rods, Spheres, and Dyads Thereof

Abstract:

Fullerene, octa(*p*-phenylene), and the dyad thereof are treated as Lennard-Jones, L-J, particles for coarse-grained molecular dynamics, MD, calculations of single-component spheres, binary rod/sphere, and ternary rod/dyad/sphere mixtures. The densities of benzene and fullerene are calculated with independently evaluated van der Waals radii. The equimolar molar rod/sphere binary mixture was found to undergo phase separation into isotropic spheres and rods in smectic phase. Doped at 0.05 mole fraction in the equimolar rod/sphere binary mixture, surfactancy is manifested by dyads preferentially locating themselves at the rod/sphere interface. It is important to note that the computed densities and phase behaviors have been verified to be independent of system size and pair potential cutoff distance over the tested ranges. The MD simulations of L-J particles accommodate both geometric effects and dispersion forces. The formation of microstructures as a result of geometric surfactancy with an increased interfacial area from the initial random distribution of rods, spheres, and dyads at constant T and P is favored by Gibbs energy depression contributed by enthalpy loss overriding the entropy loss in an L-J system. In a hard-particle system, the enthalpy loss consists only of volume reduction, which must outweigh entropy loss to support geometric surfactancy. The desired condition can be met in part by the first-order ordering transition of hard rods constituting half of the system from the initial isotropic to smectic phase accompanied by a volume reduction while hard spheres remain in the disordered state.

528404 Pd As an Oxidation State Modifier and Cocatalyst for the Re-Catalyzed Heterogeneous Deoxydehydration

Abstract:

Deoxydehydration (DODH) is an emerging biomass deoxygenation process whereby vicinal OH groups are removed. We elucidate the promotional effect of Pd for the DODH of 1,4-anhyrdoerythritol over a heterogeneous ReO_x-Pd/CeO_2 catalyst from first principles. For this mechanistic study, we performed DFT+U calculations and developed microkinetic models based on DFT data and transition state theory. Excellent agreement with experimental observations is obtained, confirming that single Re sites are the main active site for catalysis. We disentangle the contribution of the oxide support, the oxide-supported single ReO_x species, and the co-adsorbed Pd promoter that has no direct interaction with the Re species. A promotional effect of Pd on the scission of diol O-H bonds at ReO_x sites is observed. Analysis of electronic properties suggests that Pd is an oxidation state modifier of nearby Re via the CeO_2 substrate, changing the oxidation state of Re from +6 in the non-promoted catalyst to +7. Additionally, Pd serves as a catalyst for hydrogen dissociation and spillover onto CeO_2, which is an indispensable process for the regeneration of the Re catalyst. The established oxidation state modification mechanism suggests (medium-range) communication between the Pd and Re sites through the reduced ceria support.

529703 First-Principles Modeling of Single-Atom Catalysis: CO Oxidation over Atomically Dispersed Pt on CeO2

Abstract:

Supported metal catalysts are widely applied. Ceria is extensively used in exhaust gas clean-up due to the excellent oxygen storage capability and the pronounced ability to disperse the metal nanoparticles. Pt and Pd nanoparticles supported on ceria are critical in automobile three-way catalysts for pollution abatement and urban air quality control. The size of the supported metal particles is decisive for the catalytic performance. Due to the highly under-coordinated nature, single-atom catalysts (SACs) can lead to more efficient atom utilization and have the potential to enhance reactivity and selectivity. Recent advancements in experimental techniques have facilitated the synthesis and characterization of highly dispersed single metal atoms on ceria.[1-3] Single atoms on the ceria are prone to agglomeration when heated at elevated temperatures. However, surface defects of ceria are good candidates to trap Pt SAs owing to their strong interaction with low-coordination surface oxygen. The possible active sites of Pt-Ceria are still in debate. In this work, we develop a general methodology to investigate the relative stability of single-atoms on support in the presence of adsorbates at given temperature and pressure of adsorbates (e.g. CO). By combining density functional theory calculations and graph-theoretical kinetic Monte-Carlo simulations, we consider various Pt SAs and nanoparticles dispersed on ceria to investigate their CO oxidation activity, and propose that the Pt dopant has a comparable activity for CO oxidation as well as the Pt(111) facet under varying reaction conditions. The Pt atoms on terrace and steps of CeO_2 are nonactive because of the determined strong or weak CO binding, respectively. Through sensitivity analysis, the dominant surface species and the rate-determining steps are identified.

Reference:

(1) Nie, L.; Mei, D.; Xiong, H.; Peng, B.; Ren, Z.; Hernandez, X. I. P.; DeLaRiva, A.; Wang, M.; Engelhard, M. H.; Kovarik, L. ***Science*** **2017**, ***358***, 1419.

(2) Jones, J.; Xiong, H.; DeLaRiva, A. T.; Peterson, E. J.; Pham, H.; Challa, S. R.; Qi, G.; Oh, S.; Wiebenga, M. H.; Hernández, X. I. P. ***Science*** **2016**, ***353***, 150.

(3) Dvořák, F.; Camellone, M. F.; Tovt, A.; Tran, N.-D.; Negreiros, F. R.; Vorokhta, M.; Skála, T.; Matolínová, I.; Mysliveček, J.; Matolín, V. ***Nature Comm.*** **2016**, ***7,*** 10801.

530556 DFT and Microkinetic Comparison of Pt, Pd and Rh(111) for Catalytic Ammonia Oxidation

Abstract:

N_2 is the desired product over NO/N_2O for NH_3 oxidation on ammonia slip catalysts (ASC). [1] While the thermodynamic N_2 selectivity is larger than 99% at ASC conditions, the N_2 selectivity of highly active platinum group metals (PGMs) are often lower than 80%, [2] insufficient to meet the emission regulations. To date, there is no self-consistent comparison of intrinsic selectivity over PGMs. Here, we perform DFT computations of relevant reaction steps of ammonia oxidation on Pt, Pd and Rh (111). We show BrÃ¸nstedâ€“Evansâ€“Polanyi relationships for NH_x* activations by both O* and OH* and the barriers for NH_x* activation follows Pt < Pd < Rh. We build microkinetic models to predict product selectivity at reaction conditions typical of ASC and observe that the N_2 selectivity follows Pt > Rh > Pd. We observe the dependence of selectivity on *T* and pressure follows the same trend for the three metals: low *T* and $P(O_2) : P(NH_3)$ result in higher N* coverage and N_2 selectivity. The predicted selectivity of N_2, NO and N_2O agree well with experimental observations. We further perform sensitivity analysis and demonstrate the high selectivity observed for Pt originates from its smaller barriers to activate NH_x*. Higher N_2 selectivity on Rh than on Pd, however, results from the much larger barrier to form NO than N_2 on Rh, indicating $E_{a,NHx+OHx}$ and $E_{a,N2(g)\ formation}/\ E_{a,NxO(g)\ formation}$ are two important factors determining the N_2 selectivity.

REFERENCES

1. S. Shrestha, M. P. Harold, K. Kamasamudram, A. Kumar, L. Olsson, K. Leistner, Catal. Today 267(2016) 130.
2. L. Chmielarz, M. JabÅ‚onska, A. Struminski, Z. Piwowarska, A. Wegrzyn, S. Witkowski, M. Michalik, Appl. Catal. B 130-131(2013) 152.

531613 Face-Centered Tetragonal Pt Alloys of Fe & Co As Potential Catalysts for ORR

Abstract:

Proton exchange membrane fuel cells (PEMFCs) despite having advantages like high efficiency, sustainability and environment friendly operation suffer from the slow rate of oxygen reduction reaction at the cathode. The development of low-cost, efficient and stable electro-catalysts also remains a great challenge for the full-scale application of fuel cells.

In this study, we investigate special face-centered tetragonal ($L1_0$) FePt and CoPt alloys in a core-shell geometry as a potential catalyst for oxygen-reduction reactions via both dissociative and associative mechanism using density functional theory. The effect on ORR activity as a function of strain, ligand and number of Pt overlayers is studied with the help of binding energies of the reaction intermediates and full free energy landscape . A mechanistic eigenstress and eigenstrain model is used to understand the binding of a reaction intermediate as a function of strain. The ORR activity is further compared with that of Pt(111) surface.

We further show the that $L1_0$-FePt and $L1_0$-CoPt show higher bulk and surface stability than the native Pt. To enhance the stability of the Pt alloys, we look into the role of core-shell structure on surface stability as a function of Pt overlayers.

534914 Electronic Effects on Open Framework Material-Encapsulated Metal Nanoparticles (NP@OFM) and Implications on Catalysis

Abstract:

A current grand challenge of catalysis involves performing reactions with high selectivity to increase yield and efficiency of artificial catalysts in cost-effective manner. In this regard, a promising approach to impart exceptional chemoselectivity to catalytic nanoparticles (NPs) is to enshroud them with a surrounding porous shell made of open framework materials such as metal-organic frameworks (MOFs) [1]. This shell, mimicking catalytic processes in enzymes, can sterically force orientations of reactants in a particular way to make certain catalytic pathways more favorable. However, optimizing the OFM-NP combination for a desired reaction is a daunting task for experimentalists, due the millions of potential combinations to choose from and the amount of work needed to synthesize any particular combination to test it in the lab. Thus, molecular simulations are key for accelerated evaluation of possible OFM-NP combinations that can inform targets for catalyst synthesis and testing.

Toward this goal it is necessary to understand what the most crucial effects that need to be considered at the design (or computational screening) stage are. Previous studies have decoupled steric effects in NP@MOFs systems from other factors [2-4] and provided insights on how these effects can affect reactivity. However, partly due to insufficient understanding of the structure of the NP/OFM interface, the role of electronic effects originating from the interaction of OFM components and the metal at the NP/OFM interface have yet to be studied. Leveraging recent insights on the interface structure [5], here we discuss ongoing work in our lab to elucidate the relative importance of electronic effects originating from the mentioned interactions compared to electronic effects originating from changes in NP composition. We study this with the assistance of periodic DFT calculations to elucidate how charge is transferred between the OFM and the NP depending on the kind of bonds occurring at the interface, the chemical functionality of the OFM, and the composition of the metal. Thus, the effect of interactions at the interface and NP composition on the catalyst electronic density of state (DOS) and on the binding of small, common catalytic species (e.g. C*, H_3C^*, O*, HO*, N*, H*) to the NP surface has been evaluated, also considering implications on catalytic rates (e.g. via available scaling relationships).

1. P Hu, JV. Morabito, and CK Tsung. Core–Shell Catalysts of Metal Nanoparticle Core and Metal–Organic Framework Shell, ACS Catal. (2014) 4409
2. DA Gómez-Gualdrón, ST Dix, RB Getman, RQ Snurr, A modelling approach for MOF-encapsulated metal catalysts and application to n-butane oxidation Phys. Chem. Chem. Phys. (2015) 17 27596
3. ST Dix, DA Gómez-Gualdrón, RB Getman, Implications of sterically constrained n-butane oxidation reactions on the reaction mechanism and selectivity to 1-butanol, Surf. Sci. (2016) 653, 11
4. J Zhu, RB Getman, Reaction Pathways and Microkinetic Modeling of n-Butane Oxidation to 1-Butanol on Cu, Cu3Pd, Pd, Ag3Pd and PdZn (111) Surfaces, Ind. Eng. Chem. Res (2018) DOI: 10.1021/acs.iecr.8b00589
5. CL Whitford, CJ Stephenson, DA Gómez-Gualdrón, JT Hupp, OK Farha, RQ Snurr, PC Stair, Elucidating the Nanoparticle–Metal Organic Framework Interface of Pt@ ZIF-8 Catalysts J. Phys. Chem. C, (2017) 121 25079

537602 CO_2 Reduction on Ligand-Protected Au Nanoclusters

Abstract:

Although Au nanoparticles exhibit size-dependent catalytic activity, their size is usually not well defined, limiting the precise correlation of their electronic and structural properties with the observed catalysis. Atomically precise Au nanoclusters, stabilized by organic ligands, exhibit well-defined structure (size/shape), but the accessibility of reactants to the metal is limited by the presence of ligands. Experimental work has shown that these ligand-protected nanoclusters are active for CO_2 electroreduction, albeit, the active catalytic centers are elusive. In this work, we apply Density Functional Theory calculations to assess the CO_2 reduction to CO and the competing H_2 evolution reaction on thiolate protected Au nanoclusters. Our results demonstrated that the fully protected nanoclusters may not be the active catalysts for CO_2 reduction. When we accounted for the removal of a ligand from the nanocluster under electrochemical conditions, surprisingly we found that this is a thermodynamically feasible process at the experimentally applied potentials with the generated surface sites becoming active for electrocatalysis. These sites were found to significantly stabilize the COOH intermediate, resulting to dramatically facilitating CO_2 reduction. Detailed charge analysis on the nanoclusters revealed that negative charge stabilized on the nanocatalysts plays an important role in the COOH intermediate stabilization. The generated sites for CO_2 reduction are also active for H_2 evolution. This work highlights the importance of both the overall charge state and generation of catalytically active surface sites on ligand-protected nanoclusters, while elucidating the CO_2 electroreduction mechanisms. Our work rationalizes a series of experimental observations.

537727 Mechanistic Insights into Non-Oxidative Ethane Dehydrogenation on Pt-Based Catalysts Via First-Principles Microkinetic analysis

Abstract:

Mechanistic insights into non-oxidative ethane dehydrogenation on Pt-based catalysts via first-principles microkinetic analysis

Recent advancements in large-scale shale-gas production has ignited renewed interest in the use of light alkanes as low-cost feedstock for the production of chemicals and fuels. Direct dehydrogenation of ethane has been achieved at relatively low temperatures (~600°C) on Pt based catalysts. However, the efficiencies of these catalysts are still not sufficiently high due to the low yield and selectivity and quick deactivation of the catalysts under reaction conditions. A detailed mechanistic knowledge of ethane dehydrogenation reaction on the catalyst surface and the identification of the key parameters that control the selectivity and stability of these catalysts are essential in the design of more efficient catalytic materials.

We utilize a first principles based microkinetic model to develop molecular level insights into the mechanisms that control the performance and selectivity of Pt-based catalysts in ethane dehydrogenation. A thorough mechanistic study is executed on Pt(111) catalyst surface, using a network of 139 elementary steps and 55 species, to address fundamental mechanistic questions and existing debates in literature regarding ethane dehydrogenation pathways and competing reactions toward side products and coke formation. Our model successfully predicts the experimentally observed activation barrier, selectivity and partial pressure dependency of non-oxidative ethane dehydrogenation on Pt catalysts.[1,2] Sensitivity analysis identifies kinetically relevant activity and selectivity steps. The results of our analysis are used to develop a reduced reaction system and a rate expression for ethane dehydrogenation.

References:

1. Wegener, E. C. *et al.* Structure and reactivity of Ptâ€"In intermetallic alloy nanoparticles: Highly selective catalysts for ethane dehydrogenation. *Catal. Today* **299,** 146â€"153 (2018).
2. Wu, J. *et al.* Ethane and propane dehydrogenation over PtIr/Mg(Al)O. *Appl. Catal. A Gen.* **506,** 25â€"32 (2015).

538559 Predictive Model for Catalyst Effect of Photo-Induced and Copper-Catalyzed Atom Transfer Radical Polymerization (ATRP) Reaction

Abstract:

Atom transfer radical polymerization (ATRP) is among the most powerful and robust controlled radical polymerization techniques that facilitates macromolecular engineering by synthesis of polymers with precise molecular weights, low dispersities, and well-controlled architectures. Good control over polymer structures via ATRP is largely attributed to the fast activation/deactivation equilibrium that can be governed by transition metal catalysts (particularly with copper catalysts) or photoredox catalysts. However, the detailed mechanism of ATRP has not been studied thoroughly at the atomic level, which hinders the rational design of better ATRP catalysts. In this study, we employ density functional theory (DFT) and Marcus Theory calculations to investigate all possible reaction pathways for photoinduced ATRP and Copper-catalyzed ATRP (Cu-ATRP). In photoinduced ATRP, we discovered that the reaction occurs via an out-sphere electron transfer (OSET) process where the excited photoredox catalyst will be oxidized to a radical cation and the alkyl halide substrate will be reduced to radical anion that rapidly dissociate to radical for polymerization. On the other hand, inner-sphere electron transfer (ISET) pathway is most favored in Cu-ATRP in which the halogen atom in substrate transfers to the catalyst via a concerted transition state. Notably, the ISET halogen transfer transition states are first time investigated by computation in this study. For both cases, calculations provided good agreement with experimental activation rate constants. In addition, we investigated the origin of reactivity differences for a representative set of catalysts for both ATRP system by studying their electronic and steric factors. We found that for photoinduced ATRP, the E(HOMO) of catalyst can be a good predictor for the reactivity, while in Cu-ATRP, the catalyst E(HOMO), flexibility, and percentage of buried volume contribute to the reactivity deviation.

541565 25 by 25: Chemical Engineering in the Next 25 Years

Abstract:

In 2008, the centennial year for AIChE, we asked 25 chemical engineers for their visions of the profession from around the US and the world: Nine faculty, eight from industry, and eight post-docs and graduate students. We posed four questions related to each individual's industrial sector or area of research, asking for extrapolation, new impacts on existing sectors, new sectors, and an optional, more open-ended comment on the future of the profession as a whole.

In this 110th year of AIChE, we have revisited these questions with many of the same people. Ten years has seen many developments in our profession, in society, and in our thought-leaders' experiences. Here we report on their perspectives, comparing them to the 2008 visions.

541542 The Future of Chemical Engineering Itself

Abstract:

Chemical engineeringâ€™s future is bright because it is vital to so much of society and industry -- and its future is complicated by its rapidly expanding universe of activities. It must represent depth in core skills centered on the chemical sciences including biology and materials. At the same time, it must be transdisciplinary; it must exemplify the need and value of reaching across disciplines in borderless ways.

One way is to create and disseminate a compelling vision and strategy for chemical engineering, research, and education for the future. AIChE and the National Academies are working together to that end. In 1988, the National Academies published "Frontiers in Chemical Engineering: Research Needs and Opportunities.â€ Its committee of authors outlined a wide-ranging roadmap to turn promising research opportunities into reality, while calling for researchers to embrace new frontiers.

Since 1988, that has happened -- along with many other transformative changes like the Web, fracking, personalized medicine, and social media. A new nationwide effort, aided by international contributions, will address both the identity of chemical engineering and its future directions. Its goal is to ensure that the people in chemical engineering are primed to meet society's needs and rapid changes.

544708 Accelerating Innovation through Academic-Industrial Partnerships

Abstract:

In 2011, The Dow Chemical Company launched the University Partnership Initiative (UPI) to sponsor collaborative research at leading U.S. Universities. This 10-year commitment aims to address fundamental questions in industrial chemistry and chemical engineering with alignment to Dowâ€™s next-generation process technology and product development efforts. To date, this portfolio of projects has generated an impressive array of intellectual property, publications, and numerous examples of the collaborative developments implemented at Dow. This presentation will introduce UPI program and highlight several contributions from our collaborators and leaders in the chemical engineering profession.

540559 Maximizing Uptime, Efficiency, and Safety of Industrial Operations through Early Risk Detection

Abstract:

Achieving zero incidents while maximizing process uptime and efficiency is a key vision for operating companies. An important building block towards this vision is establishment of a proactive risk mitigation culture, supported by effective systems and a strong management. Studies have shown that most of the incidents and unexpected process failures can be avoided if the operating teams get timely information about developing risks and take preventive actions early on. While advances have been made in process monitoring and automation in the past decades, there still remains significant technological and behavioral gaps that prevent plant operations from proactively mitigating process risks.

A modern industrial plant monitors thousands of parameters, generating upwards of 50-100 million data points every day. Thanks to recent breakthroughs in machine learning and artificial intelligence approaches, today there are autonomous systems that can sift through this data and point out meaningful and timely insight. This can help operating teams ascertain process issues that are hidden in the data, long before process variables reach critical levels.

In this presentation, a new autonomous system, Dynamic Risk Analyzer™ (DRA), will be introduced that points out early indicators of process issues using proprietary machine learning. The risk indicators enable operations team to respond and make required process changes, days and sometimes weeks ahead of any alarm, avoiding late stage (expensive) fixes. To realize these objectives fully, a proactive management workflow must also be in place. The presentation will discuss new initiatives and workflows developed by plant operations to implement a proactive culture. Real-life case studies from plants using DRA technology will be presented on how this culture change resulted in increased uptime, efficiency and safety of plant operations.

531309 Gaussian Processes for Hybridizing Analytical & Data-Driven Decision-Making

Abstract:

Surrogate models are widely appreciated in process systems engineering [1]. The typical setting focuses on expensive-to-evaluate, possibly uncertain functions. Examples include: modular process simulators [2], integrated gasification combined cycle processes [3], a carbon capture absorber [4], and many other applications, e.g. [5-8]. Resources are typically limited, so effective decision making requires data-efficient learning.

The data science and statistical machine learning communities typically focus on models learned solely from observed data. But chemical engineering applications may also require explicit, parametric models, e.g. modeling known process constraints, operations constraints, and cost objectives [9]. So, work has integrated semi-algebraic functions with those learned from data [10] or developed semi-physical modeling techniques [11, 12].

This presentation surveys the state-of-the-art in hybridizing analytical and data-driven decision making [13]. We consider three probabilistic modeling applications to these hybrid situations:

Design of experiments for model discrimination [14]. We bridge the gap between classical, analytical methods [15] and Monte Carlo-based approaches [16]. Classical methods may have difficulty managing non-analytical model functions and data-driven Monte Carlo approaches come at a high computational cost. We replace the original, parametric models with probabilistic, non-parametric Gaussian process surrogates learned from model evaluations. The surrogates are flexible regression tools that extend classical analytical results to non-analytical models, while providing us with model prediction confidence bounds and avoiding the computational complexity of Monte-Carlo approaches.

Multi-objective optimization [17, 18]. We make novel extensions to Bayesian multi-objective optimization in the case of one analytical objective function and one black-box, i.e. simulation-based, objective function. The resulting method has been applied to a bone neotissue application [19] and a more general test suite.

Scheduling plant operations under uncertainty. For processes with equipment degradation, we use Gaussian processes to approximate large-scale, mixed-integer optimization problems.

We close by offering a broad outlook on applying probabilistic surrogate models to chemical engineering. Statistical machine learning has recently attracted significant interest in process systems engineering [20]. Here we show that state-of-the-art research in Gaussian processes [21, 22] and probabilistic modeling more generally [23] can have a big impact on chemical engineering.

References

[1] Bhosekar A, Ierapetritou M. Advances in surrogate based modeling, feasibility analysis and and optimization: A

review. *Comput Chem Eng*. 108: 250-267, **2018**.

[2] Caballero JA, Grossmann IE. An algorithm for the use of surrogate models in modular flowsheet optimization. *AIChE J*. 54(10):2633-2650, **2008**.

[3] Lang Y, Zitney SE, Biegler LT. Optimization of IGCC processes with reduced order CFD models. *Comput Chem Eng*. 35(9):1705-1717, **2011**.

[4] Cozad A, Sahinidis NV, Miller DC. Learning surrogate models for simulation-based optim-ization. *AIChE J*. 60(6):2211-2227, **2017**.

[5] Boukouvala F, Ierapetritou M. Surrogate-based optimization of expensive flowsheet modeling for continuous pharmaceutical manufacturing. *J Pharm Innov*. 8: 131, **2013**.

[6] Soepyan FB, Cremaschi S, Sarica C, and others. Estimation of percentiles using the Kriging method for uncertainty propagation. *Comput Chem Eng*. 93:143-59, **2016**.

[7] Shokry A, Ardakani MH, Escudero G, Graells M, Espuña A. Dynamic Kriging-based fault detection and diagnosis approach for nonlinear noisy dynamic processes. *Comput Chem Eng*. 106:758-76, **2017**.

[8] Tran AP, Georgakis C. On the estimation of high-dimensional surrogate models of steady-state of plant-wide processes characteristics. *Comput Chem Eng*. DOI 10.1016/j.comp-chemeng.2018.02.014, **2018**.

[9] Boukouvala F, Hasan MF, Floudas CA. Global optimization of general constrained grey-box models: new method and its application to constrained PDEs for pressure swing adsorption. *J Glob Optim*. 67(1-2): 3-42, **2017**.

[10] Boukouvala F, Floudas CA. ARGONAUT: AlgoRithms for Global Optimization of coNstrAined grey-box compUTational problems. *Optim Lett*. 11(5): 895-913, **2017**.

[11] Pearson RK, Pottmann M. Gray-box identification of block-oriented nonlinear models. *J Process Contr*. 10(4):301-315, **2000**.

[12] Cozad A, Sahinidis NV, Miller DC. A combined first-principles and data-driven approach to model building. *Comput Chem Eng*. 73: 116-27, **2015**.

[13] Boukouvala F, Misener R, Floudas CA. Global optimization advances in mixed-integer nonlinear programming, MINLP, and constrained derivative-free optimization, CDFO. *Eur J Oper Res*. 252: 701-727, **2016**.

[14] Olofsson S, Deisenroth MP, Misener R. Design of Experiments for Model Discrimination Hybridising Analytical and Data-Driven Approaches. *arXiv:1802.04170*. **2018**.

[15] Buzzi-Ferraris G, Forzatti P, Emig G, Hofmann H. Sequential experimental design for model discrimination in the case of multiple responses. *Chem Eng Sci*. 39(1): 81-85, **1984**.

[16] Vanlier J, Tiemann CA, Hilbers PAJ, van Riel NAW. Optimal experiment design for model selection in biochemical networks. *BMC Syst Biol*, 8(20), **2014**.

[17] Olofsson S, Mehrian M, Geris L, Calandra R, Deisenroth MP, Misener R. Bayesian Multi-Objective Optimisation of Neotissue Growth in a Perfusion Bioreactor Set-Up. *Computer Aided Chemical Engineering*. 40: 2155-2160, **2017**.

[18] Beykal B, Boukouvala F, Floudas CA, Pistikopoulos EN. Optimal design of energy systems using constrained grey-box multi-objective optimization. *Comput Chem Eng*. DOI 10.1016/j.compchemeng.2018.02.017, **2018**.

[19] Mehrian M, Guyot Y, Papantoniou I, Olofsson S, Sonnaert M, Misener R, Geris L. Maximizing neotissue growth kinetics in a perfusion bioreactor: An *in silico* strategy using model reduction and Bayesian optimization. *Biotechnol Bioeng*. 115(3):617-29, **2018**.

[20] Lee JH, Shin J, Realff MJ. Machine learning: Overview of the recent progresses and implications for the process systems engineering field. *Comput Chem Eng*. DOI 10.1016/j.compchemeng.2017.10.008, **2017**.

[21] Damianou A, Lawrence N. Deep Gaussian processes. In *Artificial Intelligence and Statistics*. 207-215, **2013**.

[22] Deisenroth MP, Ng JW. Distributed Gaussian processes. In *32nd International Conference on International Conference on Machine Learning*. 37: 1481-1490, **2015**.

[23] Ghahramani Z. Bayesian non-parametrics and the probabilistic approach to modelling. *Phil Trans R Soc A*. 371(1984):20110553, **2013**.

528427 Understanding the Importance of High Coverages in Electro-Catalysed Reduction of NO on Pt-Sn Alloys

Abstract:

Nitrogen cycle electrochemistry is an emerging area of interest in the chemical and environmental engineering communities, with applications ranging from removal of nitrates from wastewater streams to the development of fundamental understanding of NO electrochemistry. In spite of a significant amount of fundamental research for this chemistry on single crystal surfaces, however, the mechanistic details of the reactions are not fully known, and even basic information such as the nature of the rate-limiting step is not understood.

In this work, we begin by describing a detailed analysis of the reaction mechanism of NO electroreduction on PtSn alloys. These alloys are of interest because they have been experimentally shown to be selective to products such as hydroxylamine (NH_2OH) in acidic solutions, which is a more valuable product compared to NH_3, the only product formed on pure platinum surfaces. In spite of this interest, however, little is known about the structural and electronic features of the PtSn that underlie this selectivity.

To elucidate these features, we make use of periodic Density Function Theory calculations combined with theoretical electrochemistry analyses. We briefly review some results elucidating the mechanistic insights of NO electrochemical reduction on Pt(111)[1] and Pt(100)[2] surfaces. Next, we present our analysis for low coverages of NO present on the $Pt_3Sn(111)$ surface and derive preliminary conclusions about the reaction mechanism. Subsequently, we introduce an in-house tool to efficiently estimate higher coverage structures of adsorbed NO on the $Pt_3Sn(111)$ surface, and we extend the kinetic analysis to the limit of higher NO coverages, resembling the state of a real catalyst under continuous reaction conditions. Then, we discuss the sensitivity of the mechanistic predictions to different electrochemical reaction barrier estimation techniques, and we close by suggesting design strategies to tune the selectivity of NO electroreduction to desired products, including N_2 and hydroxylamine, on other transition metals and alloys.

References:

1. Clayborne, A., Chun, H.-J., Rankin, R. B. & Greeley, J. *Angew. Chem.* **127,** 8373â€“8376 (2015).

2. Chun, H.-J., Apaja, V., Clayborne, A., Honkala, K. & Greeley, J. *ACS Catal.* 3869â€“3882 (2017).

530270 Probing the Intrinsic Reaction Barriers of HER in Acidic and Alkaline Media Using Electronic Structure Theory

Abstract:

The hydrogen evolution reaction (HER) has been extensively studied for decades. However, several questions regarding the HER mechanism remain unanswered. In particular, the dramatic decrease in intrinsic reaction rate in alkaline media is not completely understood. Previous experimental work reported an HER activation barrier of 0.30 eV on Pt in alkaline media.[1] Recent theoretical studies argue that the activation barriers are the same in acidic and alkaline media, and that the sluggish HER kinetics in alkaline electrolytes is a consequence of an increased entropic barrier through the other Helmholtz plane.[2]

Herein, we implicitly study the effect of pH on the Volmer, Heyrovsky and Tafel steps of the HER mechanism within the density functional theory framework by changing the proton donor. We use the solvated jellium model---a method developed in our group---to study the potential-dependence of the elementary steps of the HER mechanism. This method allows for the addition/subtraction of electrons (or fractions thereof) to/from the unit cell in order to keep the potential constant.

The first elementary step of HER in any aqueous electrolyte is the Volmer step, *i.e.* proton discharge from the electrolyte to form adsorbed hydrogen. The second elementary reaction could be either the Heyrovsky or the Tafel step. The former describes hydrogen gas formation from an adsorbed hydrogen atom and a proton-electron pair from the electrolyte/circuit, whereas the latter is a purely chemical step between two adsorbed hydrogen atoms. One can thus expect the Tafel step to be independent of the applied potential save for field effects. Hence, we will report on the relative barrier heights of the Heyrovsky and Tafel steps in acidic media, and determine whether there exists a potential window within which one could expect a change in HER mechanism.

1 Sheng, W.; Gasteiger, H. A.; Shao-Horn, Y., *Journal of The Electrochemical Society*, **2010**, *157 (11)*, B1529-B1536

2 Rossmeisl, J.; Chan, K.; Skúlason, E.; Björketun, M.; Tripkovic, V., *Catalysis Today*, **2016**, *262*, 36-40

532077 First-Principles Prediction of Activated Carbon Nanostructures for Catalyzing Oxygen Reduction

Abstract:

Proton-exchange membrane (PEM) fuel cells are considered to be a promising technology for use as alternative energy sources for transport, stationary and portable applications. They use hydrogen and oxygen to generate electricity through an electrochemical process which includes the oxygen reduction reaction (ORR) at the cathode. Currently, platinum (Pt)-based catalysts are commonly used to speed up the sluggish ORR that limits the efficiency of low temperature PEM fuel cells. However, Pt is expensive and scarce, drawing much interest in developing non-Pt catalysts. Carbon nanomaterials, including fullerenes, carbon nanotubes (CNTs), graphene, and their composites, have been actively studied as a potential candidate to replace Pt-based ORR catalysts. These sp 2 -bonded carbon nanostructures are inherently constituted by largely extended Ï€-conjugation systems, making them essentially unreactive toward molecular oxygen (O_2) in their pristine forms. However, the conjugated Ï€-systems can be severely disrupted with localized Ï€ electrons through dopant incorporation, defect creation, and/or local lattice distortions. The doped, defective and distorted carbon structures are currently thought to be responsible for O_2 adsorption/activation and thus ORR, but the active sites and the exact mechanisms are still uncertain.
Using density functional theory (DFT) calculations, we have identified a novel method for the activation of graphene-like structures. This modification creates quasi-localized electron states in the Ï€-conjugated system which are vulnerable to oxidation via O_2. The result is favorable O_2 adsorption, which despite being necessary for ORR activity is not predicted for previously proposed active sites. We have identified the importance of using proper solvation models at the solid-liquid interface. While many previous DFT studies were modeled in the gas phase where O_2 adsorption is consistently predicted to be endothermic on carbon, our study clearly demonstrates that a more accurate description of interactions at the carbon-solvent interface is necessary to properly describe O_2 adsorption. In addition we have found that more detailed descriptions of charge injection are necessary at carbon-based catalysts, where polarization of the carbon electrode is significant. Ab-initio MD simulations were performed with an explicit solvent model to outline a full reaction mechanism identifying elementary reaction steps. The rate limiting step of the ORR in this system is predicted to be desorption of the final reduction products for which we predict a moderate but surmountable activation barrier. Excitingly, the predicted maximum overpotential for any elementary reaction step of the modified carbon structures that we have found is approximately 0.6 eV, which can be competitive when compared to standard carbon supported Pt (Pt/C) catalysts. Using our model of the carbon-solvent interface, we have revisited the prevalent types of activated carbon, including doping and plane edge structures, providing an improved description of the ORR activity. Our study provides insight into molecular mechanisms underlying the superior ORR activity of nanostructured carbon materials and also guidelines for the rational design and development of high-efficiency, cost-effective carbon-based ORR catalysts for fuel cell applications.

534479 Quantifying Confidence in DFT Predicted Surface Pourbaix Diagrams and Associated Reaction Pathways for Chlorine Evolution

Abstract:

Catalytic activity predictions and the identification of active sites rely on precisely determining the dominant reaction mechanism. The activity governing mechanism and products could vary with the catalyst material, which can be described by material descriptor(s), typically the binding strength(s) of key intermediate species. Density functional theory calculations can be used to identify dominant reaction mechanisms. However, the dominant reaction mechanism is sensitive to the choice of the exchange-correlation functional. Here, we demonstrate this using the chlorine evolution reaction on rutile oxides as an example, which can occur through at least three reaction mechanisms each mediated by different surface intermediates and active sites. We utilize Bayesian error estimation capabilities within the BEEF-vdW exchange-correlation (XC) functional to quantify the uncertainty associated with predictions of the operative reaction mechanism by systematically propagating the uncertainty originating from DFT-computed adsorption free energies. We construct surface Pourbaix diagrams based on the calculated adsorption free energies for rutile oxides of Ru, Ir, Ti, Pt, V, Sn and Rh. We utilize confidence-value (c-value) to determine the degree of confidence in the predicted surface phase diagrams. Using the scaling relations between the adsorption energies of intermediates we construct a generalized Pourbaix diagram showing the stable surface composition as a function of potential and the oxygen binding energy on the cus site (Î"E_O^c). This allows us to incorporate consistency between activity and surface stability, which is necessary to determine activity volcano relationships for surface reactivity. We incorporate the uncertainty in linear scaling relations to quantify the confidence in generalized Pourbaix diagram and the associated activity. This allows us to compute the expectation limiting potential as a function of Î"E_O^c , which provides a more appropriate activity measure incorporating DFT uncertainty. We show that the confidence in the classification problem of identifying the active reaction mechanism is much higher than that for the prediction problem of determining catalytic activity. We believe that such a systematic approach is needed for accurate determination of activities and reaction pathways for multi-electron electrochemical reactions such as N_2 and CO_2 reduction.

537634 A Roadmap for Modeling Single-Site (electro)Catalysts: A Combined Coupled Cluster, DFT and a Classical Force Field Approach

Abstract:

Single site catalysts supported on various substrates (e.g. graphene, carbon-nitrides, boron nitrides etc.) have recently emerged as promising catalysts for the oxygen reduction reaction (ORR). As the unique properties of these systems make them quite different from transition metals and alloy electrocatalysts, it is important to establish appropriate computational methods to accurately describe these materials. We use Cu-single atom active sites supported on a covalent triazine framework (CTF) as a case study to consider three distinct computational approaches. Using small cluster models, we first benchmark various DFT methods with coupled cluster theory to show that HSE06/D3 accurately describes the electronic structure. Next, we use hybrid DFT (HSE06/D3) calculations to determine the ORR mechanism by evaluating multiple possibilities for each reaction step. Finally, using DFT-derived force fields and classical Monte Carlo simulations, we estimate the solvation contributions using thermodynamic integration. Our results show that accurately benchmarking DFT methods (i.e. primary effects) and correctly incorporating solvation effects (i.e. secondary effects) are essential to describe the reactivity of single site (electro)catalysts. Moreover, we find that inappropriate computational methodologies may lead to the right predictions compared to experiment, but for the wrong reasons. This talk will present a broader overview of single-site ORR catalysis and summarize multiscale strategies for describing aqueous phase reactions.

PCCP

PAPER

View Article Online
View Journal | View Issue

Check for updates

Cite this: Phys. Chem. Chem. Phys., 2018, 20, 2549

The importance of grand-canonical quantum mechanical methods to describe the effect of electrode potential on the stability of intermediates involved in both electrochemical CO_2 reduction and hydrogen evolution†

Haochen Zhang,[a] William A. Goddard III,[b] Qi Lu*[a] and Mu-Jeng Cheng*[c]

Received 5th December 2017, Accepted 1st January 2018

DOI: 10.1039/c7cp08153g

rsc.li/pccp

The rational design of electrocatalysts to convert CO_2 to fuel requires predicting the effect of the electrode potential (U) on the binding and structures of the intermediates involved in CO_2 electrochemical reduction (CO_2ER). In this study, we used grand-canonical quantum mechanics (GC-QM) to keep the potential constant during the reactions (rather than keeping the charge constant as in standard QM) to investigate the effect of U on adsorption free energies (ΔGs) of 14 CO_2ER intermediates on Cu(111) as well as the intermediates involved in the competitive hydrogen evolution reaction (HER). In contrast to most previous theoretical studies where ΔGs were calculated under constant charge (= 0, neutral), we calculated ΔGs under constant potential (U = 0.0, −0.5, −1.0, and −1.5 V_{SHE}). By comparing the ΔGs calculated under constant U (= 0.0 V_{SHE}) to those calculated under constant charge, we found differences up to 0.22 eV which would change the rates at 298 K by a factor of about 5300. In particular we found that the adsorption of species with a C═O functional group (*i.e.*, *COOH, *CO, and *CHO) strengthened by up to 0.16 eV as U became more negative by 1 V, whereas the adsorption of –O– species (*i.e.*, *OH, *OCH_3, *COH, and *CHOH) weakened by up to 0.20 eV. For the (111) index surfaces of Cu, Au, Ag, Ir, Ni, Pd, Pt and Rh, we investigated the effect of U on the reaction free energy (ΔG) at pH = 0 for the crucial elementary steps for CO_2ER (*CO + (H^+/e^-) → *CHO, $\Delta G = (\Delta G_{*CHO} - \Delta G_{*CO}) + eU$) and HER (* + ($H^+/e^-$) → *H, $\Delta G = \Delta G_{*H} + eU$. Our results indicated that the influence of U on ($\Delta G_{*CHO} - \Delta G_{*CO}$) was metal dependent. In contrast, the energy for converting a proton in solution to H* on the surface, ΔG_{*H}, was barely affected by U (for the studied metals). Overall we found substantial differences (MAD > 0.18 eV) between the ΔGs calculated under U = −1.0 V_{SHE} (relevant to experiments) and those calculated under constant charge (= 0, neutral) common to most theoretical investigations. Therefore, we strongly recommend application GC-QM to obtain accurate energetics for CO_2ER.

Introduction

Recently, due to the potential to reduce the carbon footprint under mild conditions, the CO_2 electrochemical reduction (CO_2ER) has attracted much attention.[1–8] The key to the success of a commercially feasible CO_2ER process is to discover stable, efficient, and selective catalysts composed of inexpensive materials.[6] To date, only copper exhibits the ability to produce hydrocarbons and alcohols with a moderate current density (5–10 mA cm^{-2}) and efficiency (up to 69%).[9] However, poor selectivity and high overpotentials remain obstacles that prevent CO_2ER from being a commercially feasible process.[10,11]

To better understand how reactions proceed on a molecular level, the intermediates along the reaction path must be identified. In addition, the effect of the electrode potential (U) on the adsorption strength of these intermediates as well as the reaction mechanisms should be investigated. However, due to the dynamic nature of the electrocatalytic environments and complexity of the electrode–electrolyte interface, experimental studies have proven to be very challenging. Currently, only CO has been identified as an intermediate.[12–16]

In electrochemical surface science, molecular adsorption modes and adsorption free energies are U dependent.[17–27] For

[a] Department of Chemical Engineering, Tsinghua University, Beijing 100084, China. E-mail: luqicheme@mail.tsinghua.edu.cn

[b] Materials and Process Simulation Center, California Institute of Technology, Pasadena, California 91125, USA

[c] Department of Chemistry, National Cheng-Kung University, Tainan 701, Taiwan. E-mail: mjcheng@mail.ncku.edu.tw

† Electronic supplementary information (ESI) available. See DOI: 10.1039/c7cp08153g

example, Lipkowski and co-workers have studied the adsorption of pyridine at the Au(111)–water interface and determined that under negative Us, pyridine adsorbs in a flat orientation.[21] However, under positive Us, pyridine adsorbs vertically.[21] Moreover, the adsorption free energy increases from −0.52 to −0.23 eV (weaker adsorption) as U decreases from −0.06 to −0.46 V *vs.* SHE (noted as V_{SHE}). Foresti *et al.* have studied the adsorption of *n*-hexanol on Ag(111), Ag(100) and Ag(110).[24,25] On Ag(100) at $U = -0.54$ V_{SHE}, the adsorption free energy reaches a minimum value of −0.19 eV and then increases as U increases or decreases.[24,25] These experimental results along with other results indicate that the interaction between the molecular dipole moment and the local electrostatic field in the electrical double layer induced by the applied electrode potential affects the molecular adsorption mode and strength.

Recently some of us have showed how to carry out quantum mechanics (QM) calculations using grand-canonical conditions of constant potential rather than constant charge of traditional QM calculations.[28] However, this study was based on the jDFTx software, which is not generally available. In this paper we carried out grand-canonical quantum mechanics (GC-QM) using the widely available VASP software to investigate the effect of U on the adsorption strength of possible intermediates for CO_2ER and competitive HER (hydrogen evolution reaction) on Cu(111). In addition, we reported the reaction free energies of the corresponding potential-determining steps on Cu(111) and seven other transition metals. Specifically, we focused on three issues:

(a) How different are the adsorption free energies (ΔGs) on Cu(111) calculated under constant potential (const-U) from those calculated under constant charge (const-Q)?

(b) How does the U affect the ΔGs of the CO_2ER and HER intermediates on Cu(111)?

(c) How does U affect the reaction free energies of the potential-determining steps of CO_2ER and HER on the (111) index surfaces of Cu and other transition metals?

Computational details

All slab and molecular calculations were performed using the RPBE functional[29] with projector augmented wave pseudopotentials[30,31] (400 eV energy cut-off) as implemented in the Vienna *ab initio* Simulation Package (VASP).[32–35] A Gaussian-smearing technique was used with a smearing parameter of $k_BT = 0.1$ eV for slabs and 0.01 eV for molecules for the fractional occupation of the one-electron energy levels to accelerate SCF convergence. Then, all calculated electronic energies were extrapolated to $k_BT = 0$. A Monkhorst–Pack k-point net of $3 \times 3 \times 1$ was chosen to sample the reciprocal space for the slab calculations. The use of a denser $5 \times 5 \times 1$ k-point net leads to less than 0.02 eV variation in the ΔGs. For molecule calculations, only the gamma point was sampled.

The Cu(111) surface was simulated using a (3×3) symmetric five-layer slab (cell parameter = 3.615 Å for the bulk Cu) with two adsorbates on both sides of the surface, leading to

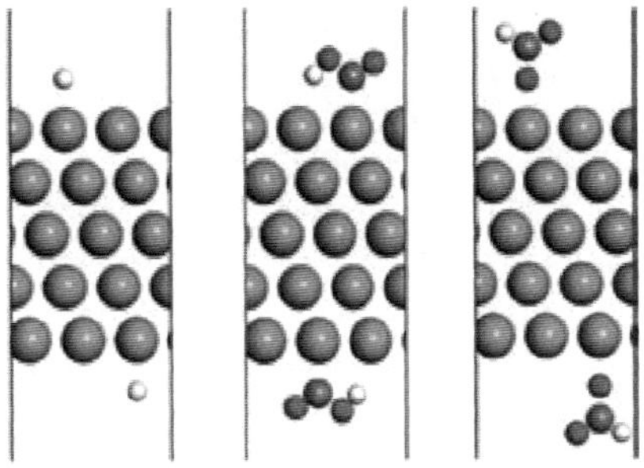

Fig. 1 Five-layer symmetrical model with *H, *COOH and *OCHO as examples.

zero dipole along the z-direction (perpendicular to the surface). During the geometric optimizations, the third layer was fixed in its bulk position, and the remaining four layers and adsorbates were allowed to relax (Fig. 1). A vacuum of at least 40 Å was introduced to avoid interactions between replicas along the z-direction. The ΔGs for adsorbed *H and *COOH intermediates (defined in the next section) calculated based on the five-layer model (0.02 and 0.78 eV for *H and *COOH, respectively) were similar to those calculated based on the seven-layer model (−0.01 and 0.75 eV for *H and *COOH, respectively), indicating that the five-layer model is sufficiently thick to represent the surface.

The free energies of the slab systems were calculated as follows:

$$G_{298\,\mathrm{K}} = E_{\mathrm{elec}}^{\mathrm{solv}} + \mathrm{ZPVE} + \sum_{\nu} \frac{h\nu}{e^{h\nu/k_BT} - 1} - T \times S_{\mathrm{vib}},$$

where $E_{\mathrm{elec}}^{\mathrm{solv}}$ is the electronic energy calculated in solvent. The vibrational frequencies were only evaluated for surface adsorbates by evaluating the partial Hessian matrix through the finite-difference approach and based on the optimized geometries under const-Q with $Q = 0$ (noted as const-Q_0). Based on the calculated vibrational frequencies, we calculated the zero-point vibrational energy (ZPVE) and vibrational contributions to the internal energy and entropy at 298 K.

The free energies of the molecules were determined as follows:

$$G_{298\,\mathrm{K}} = E_{\mathrm{elec}}^{\mathrm{solv}} + \mathrm{ZPVE} + \sum_{\nu} \frac{h\nu}{e^{h\nu/k_BT} - 1} + \frac{n}{2}k_BT - T \times (S_{\mathrm{vib}} + S_{\mathrm{rot}} + S_{\mathrm{trans}}),$$

where n is 8 for non-linear molecules and 7 for linear molecules. $E_{\mathrm{elec}}^{\mathrm{solv}}$ was obtained using the same approach as that employed for the slab systems. However, the ZPVE, the vibrational component of the internal energy, S_{vib}, S_{rot}, and S_{trans} were calculated using the PBE/6-31++G** level of theory as implemented in Jaguar. In addition, gas-phase thermochemical reaction energies that are computed with the RPBE functional exhibit systematic deviations from the experimental data.[36] Therefore, a sensitivity analysis was performed for 21 chemical reactions according to the procedure proposed by Peterson *et al.*,[37] and a correction of +0.372 eV was employed for the electronic energy of gas-phase CO_2 (see ESI†).

To treat the electrochemical interface, the approach proposed by Head-Gordon *et al.*,[38,39] Goddard *et al.*,[40,41] and Sautet *et al.*[42] was applied in this study. In this model, the Fermi energy is adjusted to a target value by changing the number of electrons in the system during each step of the geometry optimization (GC-QM), which results in the work function and electrode potential being fixed in the calculations. Then, the linear Poisson–Boltzmann implicit solvation model with a Debye screening length of 3.0 Å[43] was used to neutralize the non-zero charge in the simulation cell and simulate water and the electrolyte, allowing for a more realistic description of the electrochemical double layer. The U of the electrode was calculated by relating the work function (Φ) of the system to the experimental work function of the standard hydrogen electrode (SHE) as follows:

$$U = \frac{\Phi - \Phi_{\mathrm{SHE}}}{e}$$

A value of 4.43 eV was used for Φ_{SHE}. This number is in the range of the experimentally estimated value (*i.e.*, 4.2 ± 0.4 eV)[44] and was employed by Anderson and co-workers in their theoretical studies.[45,46] Recently, we applied this approach to study the adsorption of N-heterocyclic carbenes on Au(111).[47]

Results and discussion

ΔGs calculated under const-Q_0 and const-U_0

The major products of CO_2ER catalyzed by copper are CH_4 and C_2H_4 with oxygenates as minor products (*e.g.*, alcohols, aldehydes, ketones, and carboxylic acids).[48] In this theoretical study, we only considered the possible intermediates on the pathways to $CH_{4(g)}$ due to simplicity of the reaction mechanisms compared to those leading to the multi-carbon products.[9,49]

Numerous theoretical studies have been conducted to investigate the possible pathways for $CH_{4(g)}$ formation, as summarized in Scheme 1.[37,40,41,50] The possible intermediates include *OCHO, *COOH, *CO, *CHO, *COH, *CHOH, *C, $*OCH_3$, $*CH_2OH$, *CH, *O, $*CH_2$, *OH, and $*CH_3$, where an asterisk indicates the clean surface and adsorbates with an asterisk stands for species that are adsorbed on the surface. In most cases, the competition between HER and CO_2ER for input electricity leads to a waste of energy.[51–54] Therefore, we also calculated the ΔG for *H, which is an important intermediate in the HER.[55]

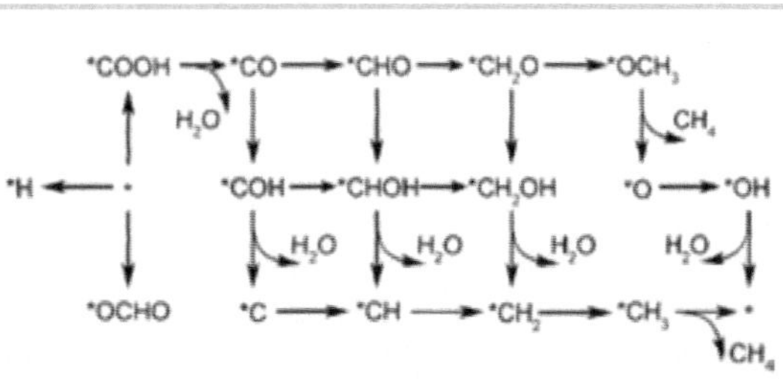

Scheme 1 Schematic representation of possible pathways for CO_2ER to $CH_{4(g)}$.

A clean Cu(111) surface, $CO_{2(g)}$, $H_2O_{(l)}$, and $H_{2(g)}$ were used as references to calculate the ΔGs of the 15 intermediates based on the following reactions:

$$* + 1/2H_{2(g)} \rightarrow *H \quad (1)$$

$$* + 1/2H_{2(g)} + CO_{2(g)} \rightarrow *COOH \quad (2)$$

$$* + 1/2H_{2(g)} + CO_{2(g)} \rightarrow *OCHO \quad (3)$$

$$* + H_{2(g)} + CO_{2(g)} \rightarrow *CO + H_2O_{(l)} \quad (4)$$

$$* + 3/2H_{2(g)} + CO_{2(g)} \rightarrow *CHO + H_2O_{(l)} \quad (5)$$

$$* + 3/2H_{2(g)} + CO_{2(g)} \rightarrow *COH + H_2O_{(l)} \quad (6)$$

$$* + 2H_{2(g)} + CO_{2(g)} \rightarrow *CHOH + H_2O_{(l)} \quad (7)$$

$$* + 2H_{2(g)} + CO_{2(g)} \rightarrow *C + 2H_2O_{(l)} \quad (8)$$

$$* + 5/2H_{2(g)} + CO_{2(g)} \rightarrow *OCH_3 + H_2O_{(l)} \quad (9)$$

$$* + 5/2H_{2(g)} + CO_{2(g)} \rightarrow *CH_2OH + H_2O_{(l)} \quad (10)$$

$$* + 5/2H_{2(g)} + CO_{2(g)} \rightarrow *CH + 2H_2O_{(l)} \quad (11)$$

$$* + 3H_{2(g)} + CO_{2(g)} \rightarrow *O + H_2O_{(l)} + CH_{4(g)} \quad (12)$$

$$* + 3H_{2(g)} + CO_{2(g)} \rightarrow *CH_2 + 2H_2O_{(l)} \quad (13)$$

$$* + 7/2H_{2(g)} + CO_{2(g)} \rightarrow *OH + H_2O_{(l)} + CH_{4(g)} \quad (14)$$

$$* + 7/2H_{2(g)} + CO_{2(g)} \rightarrow *CH_3 + 2H_2O_{(l)} \quad (15)$$

First, the ΔGs were calculated under const-Q_0. This condition has been used in most previous theoretical studies of CO_2ER catalyzed by copper.[37,50–52,56] In addition, we considered all possible adsorption sites (*e.g.*, top, bridge, fcc and hcp sites) for each intermediate. However, the reported energetics are based on the most stable site. The calculated ΔGs are summarized in Table 1. The most stable adsorption sites depend on the identity of the intermediate. The species that prefer to bind to the top site include *OCHO, *CHO, and $*CH_2OH$, and the species that prefer to bind to the three-coordinated sites (including fcc and hcp sites) include *H, *COOH, *CO, *COH, *CHOH, $*OCH_3$, *CH, *C, $*CH_3$, *O, and *OH. In addition, $*CH_2$ prefers to bind to the bridge site.

Then, the ΔGs for all 15 intermediates under const-U with $U = 0.0\ V_{\mathrm{SHE}}$ (noted as const-U_0) were calculated. In comparison to those calculated under const-Q_0, the ΔGs for *OCHO, *COOH, *CO, *CHO, *C, *CH, and *O were smaller (stronger adsorption), and the ΔGs for *H, *COH, *CHOH, $*OCH_3$, $*CH_2OH$, *OH, and $*CH_3$ were larger (weaker adsorption). However, the ΔG for $*CH_2$ did not change (Fig. 2). The mean absolute difference in the ΔGs between the two conditions ($|\Delta\Delta G| = |\Delta G(\text{const-}U_0) - \Delta G(\text{const-}Q_0)|$) was 0.10 eV with *C having the largest value (*i.e.*, 0.22 eV). The most favourable adsorption modes calculated under const-U_0 are the same as those calculated under const-Q_0 for most intermediates, except for *COOH (hcp → top), *CHOH (fcc → bridge), $*CH_3$ (fcc → top), *CH (hcp → fcc), and *H (fcc → hcp).

The major difference between const-Q_0 and const-U_0 is the change in the work function ($\Delta\Phi = \Phi_{*\mathrm{intermediate}} - \Phi_{\mathrm{Cu(111)}}$) during adsorption. $\Delta\Phi$ is not necessary zero under const-Q_0 but

Table 1 Adsorption free energies for all possible C_1 intermediates along the CO_2RR pathways calculated under constant charge (with Q = 0) and constant electrode potential (U = 0.0, −0.5, −1.0 and −1.5 V_{SHE}). The change in the work function ($\Phi_{*intermediate} - \Phi_{Cu(111)}$) during adsorption ($\Delta\Phi$) is also presented

	ΔG/eV					
Adsorbates	Const-Q_0	0.0	−0.5	−1.0	−1.5	$\Delta\Phi$/eV
*H	0.02	0.05	0.05	0.06	0.06	0.04
*COOH	0.78	0.69	0.63	0.53	0.42	−0.01
*OCHO	0.23	0.17	0.16	0.15	0.13	−0.07
*CO	0.09	−0.03	−0.07	−0.12	−0.17	0.15
*CHO	1.04	1.00	0.97	0.92	0.85	0.02
*COH	1.26	1.41	1.53	1.61	1.69	−0.39
*CHOH	1.46	1.59	1.63	1.66	1.68	−0.27
*C	2.08	1.86	1.84	1.81	1.77	0.30
*OCH_3	0.44	0.57	0.67	0.75	0.84	−0.34
*CH_2OH	1.12	1.21	1.30	1.34	1.36	−0.27
*CH	1.04	0.99	1.00	1.01	1.02	−0.01
*O	−0.47	−0.65	−0.69	−0.72	−0.75	0.22
*CH_2	0.78	0.78	0.81	0.82	0.84	−0.06
*OH	−0.81	−0.63	−0.53	−0.44	−0.34	−0.43
*CH_3	0.10	0.16	0.21	0.24	0.28	−0.43

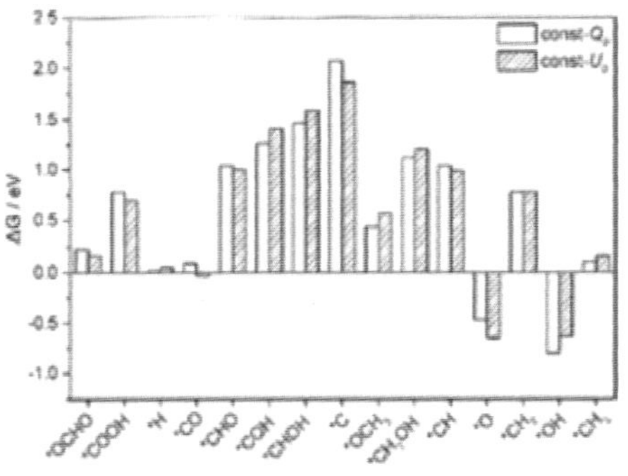

Fig. 2 Comparison of ΔGs (eV) for each intermediate calculated under const-Q_0 and const-U_0.

is precisely zero under const-U_0. Therefore, we hypothesized that $\Delta\Delta G$ is correlated to $\Delta\Phi$. Our calculated $\Delta\Phi$s are summarized in Table 1. The adsorption of *H, *CO, *CHO, *C, and *O led to an increase in Φ, whereas the adsorption of the rest of intermediates resulted in a decrease in Φ.

By comparing ΔGs and $\Delta\Phi$s, we determined that the $\Delta\Phi$s were linearly correlated to the ΔΔGs with R^2 = 0.91 (Fig. 3). As previously mentioned, the adsorption modes for five intermediates (*COOH, *CHOH, *CH_3, *CH_3 and *H) were different under const-Q_0 and const-U_0. Using ΔΔGs and $\Delta\Phi$s based on the same adsorption mode (const-U_0) improved R^2 to 0.99, which indicates that when $\Delta\Phi$ is close to zero, the ΔGs calculated under const-Q_0 are similar to those calculated under const-U_0. However, when Φ varies significantly during the process, the ΔGs calculated under the two conditions differed. Therefore, to obtain more accurate energetics, GC-QM must be employed. In addition, our study provided an equation for converting the energetics under const-Q_0 to those under const-U_0 for the Cu(111) system as follows:

$$\Delta G(\text{const-}U_0) = \Delta G(\text{const-}Q_0) - 0.543 \times \Delta\Phi - 0.054$$

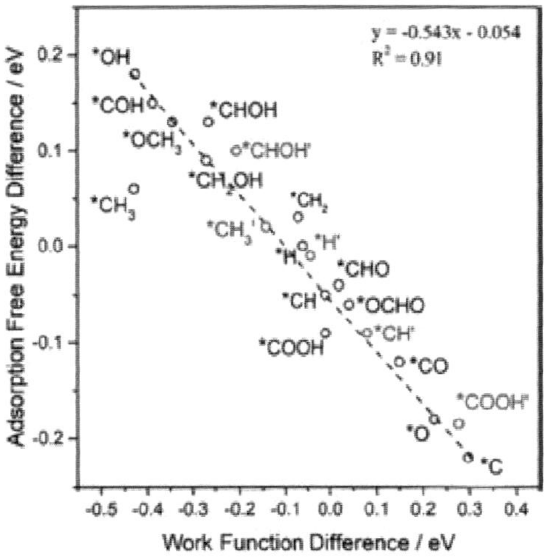

Fig. 3 Correlation between $\Delta\Phi$ and ΔΔG for 15 CO_2ER and HER intermediates. The adsorption modes for five intermediates (*COOH, *CHOH, *CH, *CH_3 and *H) were different under const-Q_0 and const-U_0. Red-coloured data points represent the situation where ΔΔGs and $\Delta\Phi$s for those intermediates were calculated based on the same adsorption mode (const-U_0).

Effect of the U on the ΔGs

For most CO_2ER catalyzed by copper, a U that is more negative than −0.5 V_{SHE} is typically required to drive the reactions in experiments.[48] Therefore, the ΔGs were calculated under more reducing conditions with U = −0.5, −1.0, and −1.5 V_{SHE} (noted as const-$U_{-0.5}$, const-$U_{-1.0}$, const-$U_{-1.5}$, respectively). A linear correlation was observed between U and ΔG for all intermediates (Table 1 and Fig. 4). For example, as U changed from 0.0 to −0.5, 1.0, and −1.5 V_{SHE}, the ΔG of *CO decreased from −0.03 to −0.07, −0.11, and −0.17 eV, respectively. Importantly, the response of ΔG to U for the individual intermediates varied differently. For example, as U decreased, the ΔGs for *COOH and *OCHO decreased but those for *COH and *CHOH increased.

Due to the linear relationship between ΔG and U for all intermediates, the data were fit to the following equation: $\Delta G = \Delta G_0 + \alpha U$. The slope ($\alpha$) represents the response of ΔG to U (Table 2), and a positive (negative) value indicates that the adsorption is stronger (weaker) as U becomes more negative. Using a α of 0.09 as the cut-off, we determined that as U became

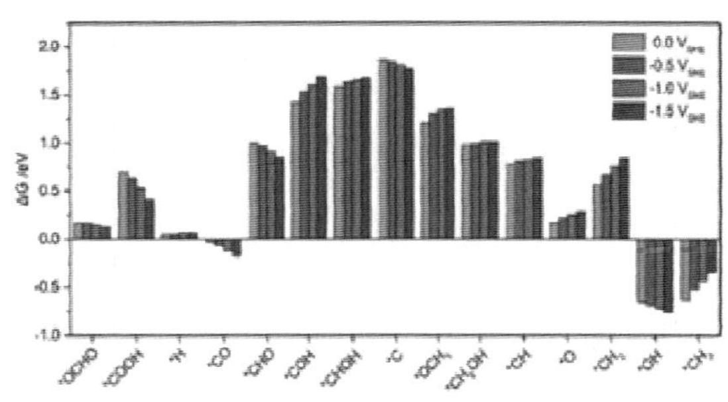

Fig. 4 ΔGs for each intermediate under U = 0.0, −0.5, −1.0 and −1.5 V_{SHE}.

more negative, the ΔGs of *COOH, *CO, and *CHO decreased. However, the ΔGs of *COH, *OCH_3, *CHOH, and *OH increased.

By comparing the functional groups of intermediates and the variations in ΔG due to U, we determined that those with a C=O motif (*COOH, *CO, and *CHO) bind to the surface more tightly as U decreased. For *CO and *CHO, this behaviour may be due to stronger π-back donation from the more negatively charged surface to the C=O motif. This argument is supported by the longer C=O bond length at more negative Us. As U changed from 0.0 to −0.5, −1.0, and −1.5 V_{SHE}, the R(C=O) (the bond length between C and O) of *CO increased from 1.192 to 1.200, 1.206, and 1.212 Å, respectively, and the R(C=O) of *CHO increased from 1.221 to 1.229, 1.233, and 1.240 Å, respectively (Scheme 2). A similar explanation has been used by Schneider to explain the variation in the CO adsorption energy on Pt(111) due to the presence of a uniform electric field.[57] For *COOH, R(C=O) remained nearly the same for the studied range of U (1.238, 1.237, 1.236, and 1.235 Å for U = 0.0, −0.5, −1.0, and −1.5 V_{SHE}, respectively). However, R(C–OH) increased from 1.365 for U = 0.0 V_{SHE} to 1.378, 1.391, and 1.403 for U = −0.5, −1.0, and −1.5 V_{SHE}, respectively.

The ΔGs for the intermediates with a -O- moiety (*OH, *OCH_3, *COH, and *CHOH) increased as U became more negative. For *OH and *OCH_3, we rationalized that this result was due to an increase in repulsion between the negatively charged surface and the lone pair of electrons on oxygen. This explanation is supported by a longer surface–adsorbate distance (for U = 0.0, −0.5, −1.0, and −1.5 V_{SHE}, these distances are 1.289, 1.305, 1.314, and 1.347 Å, respectively, for *OH and 1.375, 1.386, 1.400, and 1.412 Å, respectively, for *OCH_3). However, *COH and *CHOH are different. For these adsorbates, as U became more negative, the bond length between the surface and carbon decreased but the distance between C and O increased. As U decreased from 0.0 to −0.5, −1.0, and −1.5 V_{SHE}, the bond length between Cu(111) and C decreased from 1.253 to 1.255, 1.243, and 1.235 Å for *COH, respectively, and 1.606 to 1.596, 1.589, and 1.575 Å for *CHOH, respectively. However, the bond length between C and O increased from 1.322 to 1.330, 1.351, and 1.366 Å for *COH, respectively, and 1.357 to 1.371, 1.384, and 1.398 Å for *CHOH, respectively. This result may be due to the negative charge built up on the carbon bound to the surface, leading to increasing repulsion between C and -OH.

Nørskov and co-workers have studied the effect of uniform electric fields on the oxygen reduction reaction on Pt(111) and determined that as the field becomes more negative, the adsorption energy of *OH weakens but the adsorption energy of *O strengthens.[58] These results are similar to those observed on Cu(111) using GC-QM.

Table 2 Fitting parameters and corresponding coefficients of determination (R^2) for the response of ΔG to U

Adsorbate	ΔG_0	α	R^2
*OCHO	0.17	0.03	0.94
*COOH	0.71	0.19	0.98
*H	0.05	−0.01	0.98
*CO	−0.03	0.09	0.99
*CHO	1.01	0.10	0.97
*COH	1.42	−0.18	0.99
*CHOH	1.59	−0.06	0.96
*C	1.87	0.06	0.98
*OCH_3	0.57	−0.18	1.00
*CH_2OH	1.23	−0.10	0.89
*CH	0.99	−0.02	0.98
*O	−0.65	0.06	0.99
*CH_2	0.78	−0.04	0.98
*OH	−0.63	−0.19	1.00
*CH_3	0.17	−0.08	0.99

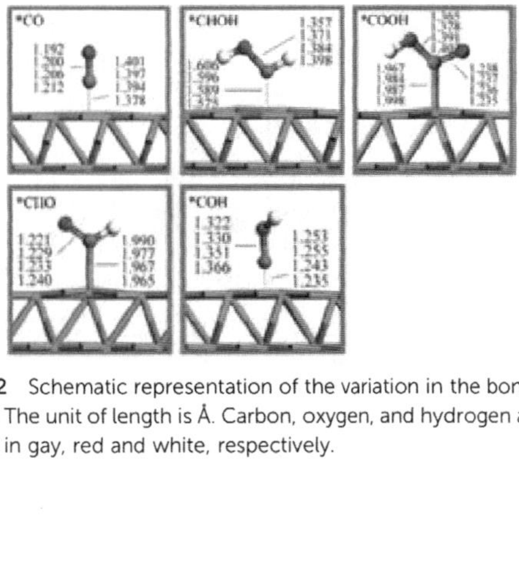

Scheme 2 Schematic representation of the variation in the bond lengths due to U. The unit of length is Å. Carbon, oxygen, and hydrogen atoms are coloured in gay, red and white, respectively.

Effect of U on the reaction free energies of the potential-determining steps of CO_2ER and HER on transition metal surfaces

Interestingly, for *CO and *CHO, the effects of U on the ΔGs were nearly identical. As U decreased from 0.0 to −0.5, −1.0, and −1.5 V_{SHE}, the ΔG of *CO decreased from −0.03 to −0.07, −0.12, and −0.17 eV, respectively, and the ΔG of *CHO decreased from 1.00 to 0.97, 0.92, and 0.85 eV, respectively. Because the ΔGs of *CO and *CHO were defined based on reactions (4) and (5), the difference between the two values is equal to the reaction free energy for the hydrogenation of *CO to *CHO using gaseous $H_{2(g)}$ (reaction (16)), and for Cu(111), this energy is nearly a constant number (*i.e.*, $\Delta G(16) = \Delta G_{*CHO} - \Delta G_{*CO} \approx 1.03$ eV) in the studied U range.

$$*CO + 1/2H_{2(g)} \rightarrow *CHO \quad (16)$$

Under electrochemical conditions, the hydrogen source is proton–electron pairs (H^+/e^-) rather than $H_{2(g)}$, and reaction (16) can be re-written as reaction (17):

$$*CO + (H^+/e^-) \rightarrow *CHO. \quad (17)$$

This elementary reaction has been proposed to be the potential-determining step for CO_2ER to CH_4 and C_2H_4 on copper.[37,56] By definition, at U = 0 and pH = 0, the Gibbs free energy of (H^+/e^-) and half that of $H_{2(g)}$ are equal ($G_{H^+/e^-} = 1/2G_{H2(g)}$), and as U changes, G_{H^+/e^-} varies correspondingly ($G_{H^+/e^-} = 1/2G_{H2(g)} - eU$).[59] As a result, $\Delta G(17) = (\Delta G_{*CHO} - \Delta G_{*CO}) + eU$ at pH = 0. By applying a negative U, the $\Delta G(17)$ can be changed from endergonic to thermal neutral or even exergonic. For Cu(111), because ($\Delta G_{*CHO} - \Delta G_{*CO}$) is U independent, $\Delta G(17)$ reaches zero when U reaches -1.03 V_{SHE}.

In addition to Cu, other metals, such as Au, Ag, Ni, Pt and Pd, exhibit the ability to reduce CO_2 to methane or methanol even though the Faraday efficiency is extremely low.[60] The $\Delta G(17)$s for the (111) index surfaces of these metals were calculated. We also considered Ir and Rh, which are common metallic centres in molecular catalysts for CO_2ER,[61–66] even though they are not active for reducing CO_2 in heterogeneous catalysis. The calculated ($\Delta G_{*CHO} - \Delta G_{*CO}$) values are summarized in Table 3.

In contrast to Cu, ($\Delta G_{*CHO} - \Delta G_{*CO}$) was U dependent for the other metals. Importantly, the response of ($\Delta G_{*CHO} - \Delta G_{*CO}$) to U was different for each metal. For Au, Ni, Pd, Pt and Rh, as U decreased from 0.0 to -1.5 V_{SHE}, ($\Delta G_{*CHO} - \Delta G_{*CO}$) increased with slopes of -0.04, -0.07, -0.15, -0.10, and -0.10 eV V_{SHE}^{-1}, respectively (Fig. 5 and Table 3). This result indicates that it becomes more difficult to make $\Delta G(17) = 0$ as U decreases, and a more negative U is required. For example, for Pd at $U = 0.0$ V_{SHE}, ($\Delta G_{*CHO} - \Delta G_{*CO}$) was 1.36 eV, and one would expect $\Delta G(17) = 0$ when $U = -1.36$ V_{SHE} is applied. However, due to the increase in ($\Delta G_{*CHO} - \Delta G_{*CO}$) as U becomes more negative, U must reach -1.61 V_{SHE} for $\Delta G(17) = 0$.

Table 3 $G_{*CHO} - \Delta G_{*CO}$) and ΔG_{*H} for all eight metals (Au, Ag, Cu, Ni, Pt, Pd, Ir and Rh) calculated under constant electric potential (U = 0.0, −0.5, −1.0 and −1.5 V_{SHE})

	($\Delta G_{*CHO} - \Delta G_{*CO}$)/eV				ΔG_{*H}/eV			
Metals	0.0	−0.5	−1.0	−1.5	0.0	−0.5	−1.0	−1.5
Ag	0.59	0.54	0.49	0.43	0.43	0.41	0.41	0.40
Au	0.51	0.54	0.56	0.57	0.44	0.44	0.45	0.46
Ir	1.02	0.98	0.95	0.88	−0.17	−0.11	−0.11	−0.11
Ni	1.56	1.61	1.64	1.67	−0.29	−0.30	−0.26	−0.25
Pd	1.36	1.44	1.52	1.59	−0.26	−0.26	−0.25	−0.29
Pt	1.13	1.19	1.24	1.29	−0.17	−0.15	−0.14	−0.11
Rh	1.24	1.31	1.36	1.40	−0.19	−0.24	−0.24	−0.23
Cu	1.03	1.03	1.03	1.02	0.05	0.05	0.06	0.06

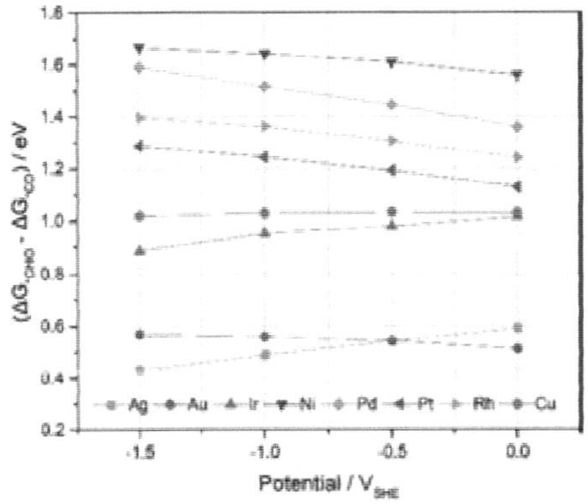

Fig. 5 $\Delta G_{*CHO} - \Delta G_{*CO}$ for the hydrogenation of *CO to *CHO for eight metals under negative potentials.

For Ag and Ir, as U becomes more negative, ($\Delta G_{*CHO} - \Delta G_{*CO}$) became smaller with slopes of 0.11 and 0.09 eV V_{SHE}^{-1}, respectively. This result indicates that as U decreases, *CO can be more easily hydrogenated to *CHO, and a less negative U is sufficient to achieve $\Delta G(17) = 0$. For example, for Ag at $U = 0.0$ V_{SHE}, ($\Delta G_{*CHO} - \Delta G_{*CO}$) was 0.59 eV, and one would expect that $\Delta G(17)$ reaches zero when $U = -0.59$ V_{SHE} is applied to the system. However, due to the decrease in ($\Delta G_{*CHO} - \Delta G_{*CO}$) as U becomes more negative, only $U = -0.54$ V_{SHE} is required to make $\Delta G(17) = 0$.

The influence of U on the reaction free energy of the two elementary reactions for HER was investigated. Replacing $H_{2(g)}$ in reaction (1) with proton–electron pairs leads to reaction (18).

$$* + (H^+/e^-) \rightarrow *H \quad \text{with } \Delta G(18) = \Delta G_{*H} + eU \qquad (18)$$

This reaction is one of the two elementary reactions in the electrochemical HER.[55] The other reaction is as follows:

$$*H + (H^+/e^-) \rightarrow * + H_{2(g)} \quad \text{with } \Delta G(19) = -\Delta G_{*H} + eU. \qquad (19)$$

Both $\Delta G(18)$ and $\Delta G(19)$ involve ΔG_{*H}, and therefore, this term has been used as a descriptor to determine the catalytic activity of materials in previous theoretical high-throughput screenings for more efficient HER electrocatalysts.[55]

In contrast to ($\Delta G_{*CHO} - \Delta G_{*CO}$), the influence of U on ΔG_{*H} was not significant for the studied metals (Fig. 6 and Table 3). The slopes were −0.01, −0.01, 0.02, −0.03, −0.03, 0.02, −0.03, and 0.02 eV V_{SHE}^{-1} for Cu Au, Ag, Ir, Ni, Pd, Pt and Rh, respectively. The lack of a functional group on *H that can interact with the induced electrostatic field leads to the insensitivity of ΔG_{H*} to U.

Comparison of energetics calculated under const-Q_0 and const-$U_{-1.0}$

In experiments, CO_2ER is conducted under const-$U_{-1.0}$.[67–70] However, in most theoretical studies, the energetics are obtained under const-Q_0.[37,50–52,56] Therefore, the energetics obtained under both conditions should be compared. The ΔGs for 15 intermediates on Cu(111) as well as ($\Delta G_{*CHO} - \Delta G_{*CO}$) and ΔG_{*H} for all eight metals under both conditions were compared. A significant difference was observed between the energetics calculated under both conditions (Fig. 7). The absolute mean

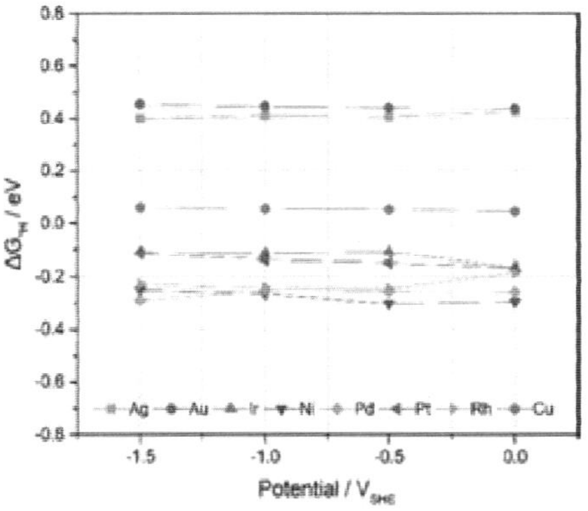

Fig. 6 ΔG_{*H} for seven metals under U = 0.0, −0.5, −1.0, and −1.5 V_{SHE}.

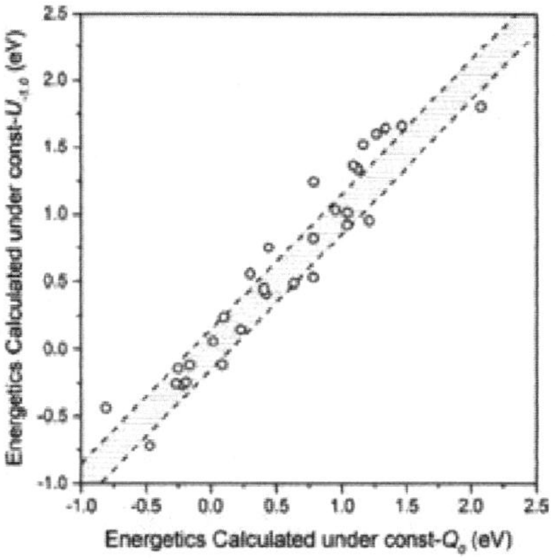

Fig. 7 Comparison of energetics calculated under const-Q_0 (standard QM) and Grand Canonical (const-$U_{-1.0}$) conditions. The green-shaded area indicates that the absolute energetic difference calculated under the two conditions is less than 0.15 eV. This may not seem large, but in factor such a chance could change the rate by a factor of 344 at 298 K, which is quite a significant. 50% of the data points lie outside of the green zone, and only half of the data points are completely inside the area, indicating that the difference between the energetics calculated under the two conditions is significant.

difference was 0.18 eV, and 15 out of 30 data points have absolute differences larger than 0.15 eV. Therefore, the use of GC-QM is important for studying CO_2ER.

Conclusions

In summary, we employed grand-canonical quantum mechanics (GC-QM) to investigate the adsorption of 15 CO_2ER and HER intermediates on Cu(111) under a constant electrode potential (U = 0.0, −0.5, −1.0, and −1.5 V_{SHE}). Unlike keeping the charge constant as in standard QM, electrochemical potential of the system is maintained unchanged by applying GC-QM. We determined that significant differences exist between the adsorption free energies (ΔGs) calculated under GC-QM or constant U = 0.0 V_{SHE} and those calculated under standard QM or constant charge (Q = 0, neutral condition). Further analysis indicated that the difference between the ΔGs calculated under the two conditions is linearly correlated to the change in the work function ($\Delta\Phi$) during the adsorption (R^2 = 0.91). This result suggests that GC-QM must be employed when $\Delta\Phi$ deviates significantly from zero. Moreover, the ΔGs for all studied intermediates were linearly affected by U. However, the response of ΔG to U varies for each intermediate. In general, for the species with a C=O functional group (*i.e.*, *CO, *CHO, and *COOH), the adsorption became stronger when U was more negative, but the opposite adsorption behaviour was observed for species with a -O- motif (*i.e.*, *OH, *OCH_3, *COH, and *CHOH).

The effect of U on the reaction free energies (ΔGs) of important elementary steps for the CO_2ER (*CO + (H^+/e^-) → *CHO, $\Delta G = (\Delta G_{*CHO} - \Delta G_{*CO}) + eU$) and HER (* + ($H^+/e^-$) → *H, $\Delta G = \Delta G_{*H} + eU$) was investigated on the (111) index surfaces of Cu, Au, Ag, Ir, Ni, Pd, Pt and Rh at pH = 0. We determined that the influence of U on the ($\Delta G_{*CHO} - \Delta G_{*CO}$) was dependent on the metal. For Au, Ni, Pd, Pt, and Rh, as U became more negative, ($\Delta G_{*CHO} - \Delta G_{*CO}$) increased, and therefore, an even more negative U was required for ΔG = 0. However, the opposite was true for Ag and Ir. For Cu, ($\Delta G_{*CHO} - \Delta G_{*CO}$) was not affected by U. In addition, for * + (H^+/e^-) → *H, ΔG_{*H} was not significantly affected by U for the studied metals. By comparing all energetics calculated under const-$U_{-1.0}$ (typical experimental condition) and const-Q_0 (commonly used in theoretical studies), we conclude that using standard QM can lead to significant differences. At 298 K, mean discrepancy of 0.18 eV would change the rates by factors up to 1107 times larger or small, which could dramatically affect the choice of optimum new catalysts. Therefore, we strongly recommend use of GC-QM for studies of CO_2ER.

Conflicts of interest

There are no conflicts to declare.

Acknowledgements

H. C. Z. and Q. L. acknowledge financial support from the National Natural Science Foundation of China (grant number 21606142). M. J. C. acknowledges financial support from the Ministry of Science and Technology of the Republic of China under grant no. MOST 105-2113-M-006-017-MY2. W. A. G. acknowledges financial support from the Joint Center for Artificial Photosynthesis, a DOE Energy Innovation Hub, supported through the Office of Science of the U.S. Department of Energy under Award No. DE-SC0004993. We wish to thank the supercomputer centre at the Tsinghua National Laboratory for Information Science and Technology for providing computational resources.

Notes and references

1 C. Costentin, M. Robert and J.-M. Savéant, *Chem. Soc. Rev.*, 2013, **42**, 2423–2436.
2 A. M. Appel, J. E. Bercaw, A. B. Bocarsly, H. Dobbek, D. L. DuBois, M. Dupuis, J. G. Ferry, E. Fujita, R. Hille and P. J. Kenis, *Chem. Rev.*, 2013, **113**, 6621–6658.
3 J. Qiao, Y. Liu, F. Hong and J. Zhang, *Chem. Soc. Rev.*, 2014, **43**, 631–675.
4 J. Albo, M. Alvarez-Guerra, P. Castaño and A. Irabien, *Green Chem.*, 2015, **17**, 2304–2324.
5 H. Takeda, C. Cometto, O. Ishitani and M. Robert, *ACS Catal.*, 2016, 70–88, DOI: 10.1021/acscatal.6b02181.
6 J. H. Montoya, L. C. Seitz, P. Chakthranont, A. Vojvodic, T. F. Jaramillo and J. K. Nørskov, *Nat. Mater.*, 2017, **16**, 70–81.
7 M. Rakowski Dubois and D. L. Dubois, *Acc. Chem. Res.*, 2009, **42**, 1974–1982.
8 A. J. Morris, G. J. Meyer and E. Fujita, *Acc. Chem. Res.*, 2009, **42**, 1983–1994.

9 R. Kortlever, J. Shen, K. J. Schouten, F. Calle-Vallejo and M. T. Koper, *J. Phys. Chem. Lett.*, 2015, **6**, 4073–4082.
10 Y. Hori, K. Kikuchi and S. Suzuki, *Chem. Lett.*, 1985, 1695–1698.
11 Y. Hori, H. Wakebe, T. Tsukamoto and O. Koga, *Electrochim. Acta*, 1994, **39**, 1833–1839.
12 A. Wuttig, C. Liu, Q. Peng, M. Yaguchi, C. H. Hendon, K. Motobayashi, S. Ye, M. Osawa and Y. Surendranath, *ACS Cent. Sci.*, 2016, **2**, 522–528.
13 J. Heyes, M. Dunwell and B. Xu, *J. Phys. Chem. C*, 2016, **120**, 17334–17341.
14 Y. Hori, A. Murata, T. Tsukamoto, H. Wakebe, O. Koga and H. Yamazaki, *Electrochim. Acta*, 1994, **39**, 2495–2500.
15 Y. Hori, O. Koga, H. Yamazaki and T. Matsuo, *Electrochim. Acta*, 1995, **40**, 2617–2622.
16 I. Oda, H. Ogasawara and M. Ito, *Langmuir*, 1996, **12**, 1094–1097.
17 C. W. T. Yang, I. Martens, E. D. L. Gyenge, R. F. Turner and D. Bizzotto, *J. Phys. Chem. C*, 2016, **120**, 2675–2683.
18 Z. Shi and J. Lipkowski, *J. Electroanal. Chem.*, 1996, **403**, 225–239.
19 L. Stolberg, J. Lipkowski and D. Irish, *J. Electroanal. Chem. Interfacial Electrochem.*, 1990, **296**, 171–189.
20 N. Li, V. Zamlynny, J. Lipkowski, F. Henglein and B. Pettinger, *J. Electroanal. Chem.*, 2002, **524**, 43–53.
21 L. Stolberg, J. Richer, J. Lipkowski and D. Irish, *J. Electroanal. Chem. Interfacial Electrochem.*, 1986, **207**, 213–234.
22 L. Stolberg, J. Lipkowski and D. Irish, *J. Electroanal. Chem. Interfacial Electrochem.*, 1987, **238**, 333–353.
23 W. B. Cai, L.-J. Wan, H. Noda, Y. Hibino, K. Ataka and M. Osawa, *Langmuir*, 1998, **14**, 6992–6998.
24 M. L. Foresti, M. Innocenti and A. Hamelin, *Langmuir*, 1995, **11**, 498–505.
25 M. L. Foresti, M. Innocenti and R. Guidelli, *J. Electroanal. Chem.*, 1994, **376**, 85–95.
26 A. Iannelli, J. Merza and J. Lipkowski, *J. Electroanal. Chem.*, 1994, **376**, 49–57.
27 D. F. Yang and J. Lipkowski, *Langmuir*, 1994, **10**, 2647–2653.
28 R. Sundararaman, W. A. Goddard and T. A. Arias, *J. Chem. Phys.*, 2017, **146**, 114104.
29 B. Hammer, L. B. Hansen and J. K. Nørskov, *Phys. Rev. B: Condens. Matter Mater. Phys.*, 1999, **59**, 7413.
30 G. Kresse and D. Joubert, *Phys. Rev. B: Condens. Matter Mater. Phys.*, 1999, **59**, 1758.
31 P. E. Blöchl, *Phys. Rev. B: Condens. Matter Mater. Phys.*, 1994, **50**, 17953.
32 G. Kresse and J. Hafner, *Phys. Rev. B: Condens. Matter Mater. Phys.*, 1994, **49**, 14251.
33 G. Kresse and J. Hafner, *Phys. Rev. B: Condens. Matter Mater. Phys.*, 1993, **47**, 558.
34 G. Kresse and J. Furthmüller, *Comput. Mater. Sci.*, 1996, **6**, 15–50.
35 G. Kresse and J. Furthmüller, *Phys. Rev. B: Condens. Matter Mater. Phys.*, 1996, **54**, 11169.
36 D. W. Blaylock, T. Ogura, W. H. Green and G. J. Beran, *J. Phys. Chem. C*, 2009, **113**, 4898–4908.
37 A. A. Peterson, F. Abild-Pedersen, F. Studt, J. Rossmeisl and J. K. Nørskov, *Energy Environ. Sci.*, 2010, **3**, 1311.
38 J. D. Goodpaster, A. T. Bell and M. Head-Gordon, *J. Phys. Chem. Lett.*, 2016, **7**, 1471–1477.
39 M. R. Singh, J. D. Goodpaster, A. Z. Weber, M. Head-Gordon and A. T. Bell, *Proc. Natl. Acad. Sci. U. S. A.*, 2017, **114**, E8812–E8821.
40 H. Xiao, T. Cheng and W. A. Goddard, *J. Am. Chem. Soc.*, 2016, **139**, 130–136.
41 H. Xiao, T. Cheng, W. A. Goddard and R. Sundararaman, *J. Am. Chem. Soc.*, 2016, **138**, 483–486.
42 S. N. Steinmann and P. Sautet, *J. Phys. Chem. C*, 2016, **120**, 5619–5623.
43 K. Mathew and R. G. Hennig, 2016, arXiv preprint arXiv: 1601.03346.
44 W. A. Donald, R. D. Leib, J. T. O'Brien, M. F. Bush and E. R. Williams, *J. Am. Chem. Soc.*, 2008, **130**, 3371–3381.
45 R. Jinnouchi and A. B. Anderson, *J. Phys. Chem. C*, 2008, **112**, 8747–8750.
46 R. Jinnouchi and A. B. Anderson, *Phys. Rev. B: Condens. Matter Mater. Phys.*, 2008, **77**, 245417.
47 K. Chang, J. G. Chen, Q. Lu and M. J. Cheng, *J. Phys. Chem. A*, 2017, **121**, 2674–2682.
48 K. P. Kuhl, E. R. Cave, D. N. Abram and T. F. Jaramillo, *Energy Environ. Sci.*, 2012, **5**, 7050.
49 W. Luo, X. Nie, M. J. Janik and A. Asthagiri, *ACS Catal.*, 2015, **6**, 219–229.
50 X. Nie, M. R. Esopi, M. J. Janik and A. Asthagiri, *Angew. Chem., Int. Ed.*, 2013, **52**, 2459–2462.
51 M.-J. Cheng, E. L. Clark, H. H. Pham, A. T. Bell and M. Head-Gordon, *ACS Catal.*, 2016, **6**, 7769–7777.
52 M.-J. Cheng, Y. Kwon, M. Head-Gordon and A. T. Bell, *J. Phys. Chem. C*, 2015, **119**, 21345–21352.
53 M. Karamad, V. Tripkovic and J. Rossmeisl, *ACS Catal.*, 2014, **4**, 2268–2273.
54 V. Tripkovic, M. Vanin, M. Karamad, M. E. Björketun, K. W. Jacobsen, K. S. Thygesen and J. Rossmeisl, *J. Phys. Chem. C*, 2013, **117**, 9187–9195.
55 J. Greeley, T. F. Jaramillo, J. Bonde, I. Chorkendorff and J. K. Nørskov, *Nat. Mater.*, 2006, **5**, 909–913.
56 A. A. Peterson and J. K. Nørskov, *J. Phys. Chem. Lett.*, 2012, **3**, 251–258.
57 P. Deshlahra, E. Wolf and W. Schneider, *J. Phys. Chem. A*, 2009, **113**, 4125–4133.
58 G. Karlberg, J. Rossmeisl and J. K. Nørskov, *Phys. Chem. Chem. Phys.*, 2007, **9**, 5158–5161.
59 J. K. Nørskov, J. Rossmeisl, A. Logadottir, L. Lindqvist, J. R. Kitchin, T. Bligaard and H. Jonsson, *J. Phys. Chem. B*, 2004, **108**, 17886–17892.
60 K. P. Kuhl, T. Hatsukade, E. R. Cave, D. N. Abram, J. Kibsgaard and T. F. Jaramillo, *J. Am. Chem. Soc.*, 2014, **136**, 14107–14113.
61 I. Osadchuk, T. Tamm and M. R. S. Ahlquist, *ACS Catal.*, 2016, **6**, 3834–3839.
62 C. Caix, S. Chardon-Noblat and A. Deronzier, *J. Electroanal. Chem.*, 1997, **434**, 163–170.
63 P. Kang, T. J. Meyer and M. Brookhart, *Chem. Sci.*, 2013, **4**, 3497–3502.

64 P. Kang, S. Zhang, T. J. Meyer and M. Brookhart, *Angew. Chem., Int. Ed.*, 2014, **53**, 8709–8713.

65 S. T. Ahn, E. A. Bielinski, E. M. Lane, Y. Chen, W. H. Bernskoetter, N. Hazari and G. T. R. Palmore, *Chem. Commun.*, 2015, **51**, 5947–5950.

66 P. Kang, C. Cheng, Z. Chen, C. K. Schauer, T. J. Meyer and M. Brookhart, *J. Am. Chem. Soc.*, 2012, **134**, 5500–5503.

67 Z. Han, R. Kortlever, H.-Y. Chen, J. C. Peters and T. Agapie, *ACS Cent. Sci.*, 2017, **3**, 853–859.

68 C. Hahn, T. Hatsukade, Y.-G. Kim, A. Vailionis, J. H. Baricuatro, D. C. Higgins, S. A. Nitopi, M. P. Soriaga and T. F. Jaramillo, *Proc. Natl. Acad. Sci. U. S. A.*, 2017, 201618935.

69 Y. Kwon, Y. Lum, E. L. Clark, J. W. Ager and A. T. Bell, *ChemElectroChem*, 2016, **3**, 1012–1019.

70 A. Loiudice, P. Lobaccaro, E. A. Kamali, T. Thao, B. H. Huang, J. W. Ager and R. Buonsanti, *Angew. Chem., Int. Ed.*, 2016, **55**, 5789–5792.

539489 Trends in Electrochemical Oxygen Reduction and Evolution Activities of Layered Double Hydroxides

Abstract:

Layered double hydroxides (LDH) have a general stoichiometry of A_xMO_2, where M is a first-row transition metal and A is an alkali intercalated metal or a proton. LDHs were originally discovered as battery electrode materials, but presently are also the most active oxygen evolution reaction (OER) catalysts in alkaline media [1]. However, their activities for OER and for oxygen reduction reaction (ORR) in full range of $A_x$and M stoichiometry is largely unexplored.

In this work, we perform a systematic high-throughput screening of $A_xMO_2$materials as bifunctional catalysts for OER and ORR using a simplified model. Our model is based on observation that the electro-catalytic activity to a large degree is defined by the surface coordination of the active site (M-atom) and is almost independent of the type of the A-atom or the layer-to-layer spacing. Using this model, we have found that our screening correctly detects the known active LDHs, but also identifies new very active OER and also ORR catalysts. The obtained systematic trends will also help to establish a principle to rationally design other materials as bifunctional catalysts for OER and ORR.

[1] S. Barwe, C. Andronescu, J. Masa, and W. Schuhmann, Current Opinion in Electrochemistry (2017).

541411 Programmable Assembly of Multiflavored DNA-Functionalized Particles

Abstract:

In this talk, I will describe recent work from my group in which we leverage self-assembly of DNA-functionalized particles (DFPs) for the design of binary superlattices. In conventional binary superlattice formation, particle size asymmetry is often the most important control variable which allows for the creation of lattices with varying component stoichiometries such as AB, AB2, AB3, etc. I will discuss our synergistic computational and experimental efforts to remove this constraint on fine tunability of particle size modality, and instead demonstrate how flexibility afforded by DNA-mediated interactions can be used to tune interparticle interactions between particles of the same size but separately tailored function to assemble particles into various crystalline morphologies. In the two-dimensional case, we demonstrate the ability to dial in a wide range of structures, including the formation of checkerboard (square), alternating string (hexagonal), honeycomb (hexagonal), Kagome and square Kagome (hexagonal) lattices. Beyond tailoring the 'flavor' of interparticle interactions, I will also discuss the formation of open lattices by tuning the range of interparticle interactions, which opens new avenues for the design of hierarchically structured and crystalline porous materials through long-sought after bottom-up strategies for building block assembly.

541811 Coarse-Grained Models for Understanding Transport in Soft, Crowded Matrices

Abstract:

Within biological systems, macromolecular crowding affects the structure, assembly, and dynamics of proteins, biopolymers, and other nanoscopic constituents. Understanding these effects provides insight into biological function and enables new strategies for delivering drugs and theranostics. The transport processes of nanoscopic constituents are well-described by continuum theories when their size is much larger or much smaller than typical length scales within the crowded matrix. The continuum assumption breaks down, however, when the transported constituents are comparable in size to features of the matrix. In this limit, transport remains incompletely understood. In my talk, I will discuss our recent application of coarse-grained models to investigate particle transport in soft polymer matrices, as simplified models for crowded media. We use these models to elucidate the physical mechanisms responsible for the anomalous transport behavior that has been recently reported in complementary microscopy experiments. Our results show that these unusual dynamics arise due to the coupling of particle motions with comparably sized fluctuations in the surrounding matrix.

541812 Engineering Pathways across Biological Barriers

Abstract:

Engineering pathways across biological barriers is entering a new era with the rapid advancement of computational resources. Our group focuses on developing multiscale simulation methods (atomistic and coarse-grained) to elucidate the interfacial phenomena associated with biological barriers that play a role in life-threatening diseases such as Alzheimerâ€™s disease, cancer, and chronic infections. In the past few years, we have made breakthroughs in three research domains - elucidated the molecular architecture of the blood-brain barrier and developed strategies to enhance the barrierâ€™s permeability for treatment of the neurodegenerative diseases; developed telodendrimer based nanocarriers for efficient delivery of approved anticancer drugs for treatment of solid tumors; and designed an online computational platform to screen libraries of small molecules for their permeability across bacterial membranes and determine their use as antibiotics for treatment of chronic infections. Understanding the molecular basis of these biological barriers has been difficult using experimental methods. Our goal is to influence the experimentally dominated research fields by providing mechanistic, structural, and molecular insights into the biological barriers that were computationally unattainable just a few years ago.

541412 Methodologies for Enhanced Unbiased Sampling of the Free Energy Landscapes of Proteins

Abstract:

Protein function is inextricably linked to protein dynamics. As we move from a static structural picture to a dynamic ensemble view of protein structure and function, novel computational paradigms are required for observing and understanding conformational dynamics of proteins and its functional implications. In principle, molecular dynamics simulations can provide the time evolution of atomistic models of proteins, but the long time scales associated with functional dynamics make it difficult to observe rare dynamical transitions on the free energy landscapes associated with protein function. One approach to surmount this challenge is to use reaction coordinate that characterize the rare conformational transitions to sample conformational landscapes more effectively. However, it is not possible to identify reaction coordinate for systems a priori, especially for proteins with limited availability of structural information. Furthermore, different reaction coordinates characterize the free energy barriers at different points on the free energy landscapes. Here, we discuss a set of methodologies outlined in a series of papers from our group, which use evolutionarily coupling between protein residues and machine learning methods to adaptively guide sampling with no alterations to potential energy functions in order to expedite exploration of the free energy landscapes.

541978 Taming the Complexity of a Messy Solution Process to Find Better Materials for Solar Cells: A Bayesian Optimization-Guided Route to Better Bouillabaisse

Abstract:

The nucleation of hybrid organic-inorganic perovskites (HOIPs) in solution occurs by a molecular self-assembly process that remains poorly understood and under-investigated. Despite progress to date, the lack of understanding of processing-structure-property relationships in HOIPs inhibits the community from solving several key challenges that could, for example, mitigate issues currently surrounding scale-up of these photovoltaic (PV) devices or their stability. There is considerable current interest in having a fundamental understanding of the interactions that take place between the constituents of the medium from which HOIPs will ultimately nucleate and grow. This puts a premium on understanding molecular-scale interactions between the choices of solvent(s), lead salt, and organic and/or inorganic cations that will form a particular HOIP crystal. We tackle this complexity in two ways: Firstly, a detailed quantum mechanical level study of the interactions between the species in the process, our "bouillabaisse," and secondly, using Bayesian optimization techniques to vastly accelerate the search for the combinations of chemical species in the "soup" that lead to a desired performance outcome, e.g., best binding energy between chosen species. This study includes investigating "solvent engineering," namely, assessing the interaction of lead salts and ancillary cationic species as we change processing solvents (both "bath" solvents and "anti-solvents") and determine the impact of solvent choice on complexation and formation of perovskite crystal building blocks. We will show how the solvating strength of the bath solvents, while helpful in solubilizing the lead, hinders the formation of perovskite-generating nuclei and prohibits precipitation of crystallites in solution. We observe that coordination of the solvent with the Pb atom causes a change in Pb-halide bond length and the shift in electron density prohibits complexation with the cation. Our *ab initio* calculations show that the binding energy between the two major building blocks of the final perovskite crystal structure can be increased using an anti-solvent to tune the overall relative dielectric of the solvent. As the relative dielectric of the solution decreases, the distance between lead and halide ions, and that between halide and cation also (favorably) decrease. This provided an important new way to tune the crystallization outcome.

535411 The Properties of Gases and Liquids: 2020

Abstract:

For decades, The Properties of Gases and Liquids (PGL) has been a standard reference for professionals in the field, advanced undergraduate, and graduate students. With 27,000 google scholar citations and counting, this book serves a similar role to "Consumer Reports" in objectively evaluating the range of estimation methods available in the literature. Nevertheless, this important resource has not been updated since 2001. The current authors have undertaken to provide this update, with careful attention to the many changes in technology and methods that have occurred over the years. Our mission is to survey the entire field of physical property estimation in order to promote the best approaches and objectively characterize the quality of the available data and methods, such that current applications are reliable and future progress of the field is advanced.

A lot has happened since the fifth edition of PGL was published. Broadly speaking, advanced computer methods have become much more feasible for practical engineering calculations. These include quantum mechanical (QM) methods like Moller-Plesset and quantum density functional theory (DFT), quantitative structure property relations (QSPR; both closed-form and machine-learning methods), and molecular simulations like Monte Carlo (MC) and molecular dynamics (MD) methods. Additionally, group contribution (GC) methods have become more sophisticated, with second and third order groups in many cases. Despite the increased sophistication, these approaches are not guaranteed to be more accurate than traditional methods, so rigorous evaluation of all available methods, both old and new, is essential for practical application. Doing so requires the specification of a common set of metrics against which the performance of all methods is compared. This set of metrics cannot simply be obtained from the original literature as the range of compounds, conditions, and comparative statistics varies from study to study. As in previous versions of PGL, care is taken in this edition to define an objective and comprehensive set of metrics for proper evaluation of existing methods so that proper recommendations may be made about predictive methods.

Modern chemical databases and computer automation make evaluation possible in a manner not available previously, but this increased capacity for objective comparisons is consistent with PGL editions of the past. For many years, the spirit of PGL has reflected the working hypothesis that predictive methods are most reliable when they combine the fundamental principles of physics as a foundation with an appropriate degree of empiricism to ensure accurate reflection of experimental measurements. In this sense, experimental measurement comprises the final arbiter of any engineering model. Therefore, the first order of business in evaluating physical property methods is the compilation of standard databases for comparing the methods on a consistent basis. Despite best intentions, it would not be feasible for these authors to code and test every method that has been published since the fifth edition. The advancing sophistication of predictive models has made it more difficult to reproduce and evaluate many methods. Therefore, we envision a two stage process for making comparisons. In the first stage, method developers will submit working executables of their method in accordance with working protocols that we define. These submissions will be evaluated against the databases and the top few methods will be selected for the sample calculation stage. In the second stage, sample calculations will be performed by the authors. Final recommendations will include the sample calculations in textbook form along with a possibly subjective consideration of how easily reproducible the calculations would be for our readers to perform.

In this presentation, we outline our expectations for database development and method comparisons in greater detail.

528118 Computer-Aided Polymer Design Using COSMO-RS

Abstract:

COSMO-RS[1] is a successful and accurate statistical thermodynamics method that has distinct advantages for computer-aided molecular design (CAMD)[2]: (1) it incorporates QM-derived electronic and structural information; (2) thermodynamic property estimation of fluids is extremely fast assuming that a unique surface charge distribution called a sigma profile is known for each species in solution; and (3) COSMO-RS as a technique is amenable to integration with common CAMD strategies[3,4]. However, CAMD using COSMO-RS is still a developing research area and has thus far typically focused on small molecule and solvent design. Addressing polymeric systems with COSMO-RS-based CAMD opens these powerful design techniques to diverse applications in membranes, electronics, aerospace materials, green chemistry, etc. Polymer design with COSMO-RS also benefits from the methodâ€™s accuracy in this domain as has recently been shown for a few properties[5].

We address the optimization of polymers in this work with two primary strategies. The first leverages the simplicity of calculating the sigma profiles of polymeric structures from monomeric sub-units[5]. This strategy naturally leads to a mixed-integer nonlinear programming (MINLP) formulation which allows for the design of a copolymer from a large number of monomeric candidates. As a second strategy, we project the high-dimensional search space of polymeric structures into a lower dimensional space of the sigma moments (the statistical moments of the sigma profiles). This allows for the use of state-of-the-art derivative-free optimization (DFO) algorithms to address the design problem in a meaningful lower-dimensional space, for which they have been shown to be effective at finding good solutions in CAMD problems[3]. Because this strategy designs structures at each iteration, we also discuss the possibility of using molecular dynamics (MD) simulations for each designed system to improve the accuracy of the property estimates. We apply the methods to a few case studies drawn from the polymer literature.

1. Klamt A, Jonas V, Bürger T, Lohrenz JC. Refinement and parametrization of COSMO-RS. The Journal of Physical Chemistry A. **1998** Jun 25;102(26):5074-85.
2. Austin ND, Sahinidis NV, Trahan DW. Computer-aided molecular design: an introduction and review of tools, applications, and solution techniques. Chemical Engineering Research and Design. **2016** Dec 1;116:2-6.
3. Austin ND, Sahinidis NV, Trahan DW. A COSMO-based approach to computer-aided mixture design. Chemical Engineering Science. **2017** Feb 23;159:93-105.
4. Scheffczyk J, Fleitmann L, Schwarz A, Lampe M, Bardow A, Leonhard K. COSMO-CAMD: A framework for optimization-based computer-aided molecular design using COSMO-RS. Chemical Engineering Science. **2017** Feb 23;159:84-92.
5. Loschen C, Klamt A. Prediction of solubilities and partition coefficients in polymers using COSMO-RS. Industrial & Engineering Chemistry Research. **2014** Jul 7;53(28):11478-87.

534840 Design of Electrosprayed Polysaccharides Nano/Microparticles for Drug and Vaccine Delivery

Abstract:

Polysaccharides are natural polymers that are attractive for a variety of applications because of their renewability and biocompatability. Particles prepared from polysaccharides are of interest as carriers for controlled release in food, cosmetics and pharmaceuticals, and as platforms for biosensing in environmental and biomedical fields. Electrohydrodynamic atomization or electrospraying is an attractive process, in terms of controllability and scalability, for synthesizing such particles. The objective of our study was to investigate the effects of individual factors in the preparation of polysaccharide particles by the electrospray process on the particle size and shape using factorial design and response surface methodology. Two types of polysaccharides, namely the anionic alginate and the cationic chitosan, were investigated. The response surface method (RSM) experiments were conducted for sodium alginate based on the results of the one-half fractional-factorial design with five factors (flowrate, voltage, polymer concentration, collection distance, and surface tension of the collection medium) and two levels (low, high). The microgel particle size of sodium alginate was found to be primarily dependent on the polymer concentration, flowrate, and applied voltage. Formation of non-spherical particles was also observed for certain molecular weights of the polymer. Similar experiments were performed with for chitosan, with molecular weight of chitosan, polymer concentration, acetic acid concentration, solution flowrate, voltage, nozzle diameter and collection distance as factors. Toward the development of a process for protein-encapsulation, electrospraying of proteinâ€"chitosan polyelectrolyte complexes were studied. Bovine serum albumin was used as a representative low-isoelectric-point protein. Split-plot RSM with seven factors was conducted to understand the effects on particle size and protein encapsulation efficiency.

References:

Myrick, JM. Stimulus-Responsive Polysaccharide Microgel Particles Prepared By Temperature-Induced Aggregation and Electrospray Techniques. Ph.D. Dissertation, Clarkson University, May 2017

Myrick JM, Vendra VK, Krishnan S. Self-Assembled Polysaccharide Nanostructures for Controlled-Release Applications. *Nanotechnol. Rev.* 2014, 3, 319â€"346, DOI: https://doi.org/10.1515/ntrev-2012-0050

Sexton FA, Krishnan S, Vendra VK. Sustained Release of Nutrients In Vivo. Patent No. US 8563066 B2 (2013)

538371 QSAR Study of Combretastatin-like Chalcones As Cancer Cell Growth Inhibitors Using Linear and Non-Linear Machine Learning Approaches

Abstract:

Combretastatin-like chalcones are known to be effective inhibitors for the mitotic process that tends to help the growth of cancer cells. Chalcones interact with ð±-tubulins, and inhibit the polymerization process that ends up increasing tumor size [1]. Based on the works of Ducki et al [2], 87 chalcones were selected. It is important to mention that an attempt to model inhibitory concentration (pIC_{50}) was made by Lipinski et al [3]. Although they have achieved models with acceptable fitness, there is still room for improvements. It can be noticed in their work that, although they used a variety of descriptor classes, their use of model development algorithms is lacking in light of recent advances.

This project aims at modifying, improving, and implementing machine learning and model development algorithms such as Artificial Bee Colony, Multi-gene Genetic Algorithm, Particle Swarm Optimization algorithms in combination with linear, non-linear, and LASSO regression algorithms. The goal is to develop models with better fitness using descriptors that are comparatively easier to calculate (e.g. connectivity descriptors, 2D descriptors). Avogadro open access software was used to develop and optimize the molecular structures, and Dragon 6.0 software was used to generate descriptors for the given compounds. One-fifth of the samples were used for external validation of model. Model superiority was decided based on values of R^2, Q^2, MSE, and MAE values of the models.

Selected References:

[1] Sharma et al, A review of mechanisms of antitumor activity of chalcones, 2016, Anticancer Agents Med Chem, 16, 200-211.

[2] Ducki et al, Combretastatin-like chalcoles as inhibitors of microtubule polymerization. Part 1: synthesis and biological evaluation of anitvascular activity, 2009, Bioorg Med Chem, 17, 7698-7710.

[3] Lippinski et al, A molecular modeling study of Combretastatin-like chalcones as anticancer agents using PLS, ANN and consensus models, 2018, Struct Chem (2018), https://doi.org/10.1007/s11224-017-1072-2

524501 Integration of Heuristic Knowledge in a Skin-Care Emulsion Design

Abstract:

The design of optimal mixtures is considered an important challenge in many industrial sectors (Cheng et al., 2009; Gani, 2004; Lee et al., 2014), mainly due to the need for a limited searching space for the most appropriate components, and their level of use within the final formulation. In recent years, many modelling approaches have been proposed to systematize and accelerate this screening process. Such approaches usually integrate rule-based elements and/or experiments in a combinatorial modelling problem with highly nonlinear expressions relating composition and product properties (Smith and Ierapepritou, 2010). Besides, in the industrial domain of cosmetic formulations, there is a considerable amount of available heuristic knowledge, regarding qualitative function of ingredients, as well as their impact on sensorial attributes. In a previous study (Arrieta-Escobar et al., 2017), a CAMD methodology (Achenie et al., 2002) integrated with heuristics was proposed for the case of hair care emulsions using mixed-integer optimization with logical constraints (Williams, 2013).

Here we illustrate the CAMD methodology integrated with heuristics and its use in the cosmetic emulsions design, using a body lotion as an example. Initially, the general structure of the methodology is presented; then, the heuristics regarding formulation and manufacture, the selected performance parameters for this kind of products, and the details of the experimental validation are described. The results show that the incorporation of heuristics into a systematic CAMD methodology using integer programming with logical constraints enables to obtain a set of feasible alternatives, which are optimised to meet the specified requirements of performance with a minimal cost. The obtained emulsion design was validated by developing and manufacturing a set of body lotion samples. The rheological, textural and microstructural properties of the samples, along with their consumer assessment, show that these alternatives met the expected properties. This methodology could be used as a rapid preliminary design tool for inexperienced cosmetic scientists and designers. Also it can help to reduce the time and resources spent in cosmetic emulsions design, comparing with the traditional trial-and-error methods.

References

Achenie, L.E.K., Gani, R., Venkatasubramanian, V. (Eds.), 2002. Computer Aided Molecular Design: Theory and Practice. Elsevier, Amsterdam.

Arrieta-Escobar, J.A., Bernardo, F.P., Orjuela, A., Camargo, M., Morel, L., 2017. An Integrated Methodology for Emulsified Cosmetic Product Formulation Using Integer Programming with Logical Constraints, in: Espuña, A., Graells, M., Puigjaner, L. (Eds.), Computer Aided Chemical Engineering. Elsevier B.V., Barcelona, Spain, pp. 985–990. https://doi.org/10.1016/B978-0-444-63965-3.50166-5

Cheng, Y.S., Lam, K.W., Ng, K.M., Ko, R.K.M., Wibowo, C., 2009. An integrative approach to product development—A skin-care cream. Comput. Chem. Eng. 33, 1097–1113.

https://doi.org/10.1016/j.compchemeng.2008.10.010

Gani, R., 2004. Chemical product design: challenges and opportunities. Comput. Chem. Eng. 28, 2441–2457. https://doi.org/10.1016/j.compchemeng.2004.08.010

Lee, C.K.H., Choy, K.L., Chan, Y.N., 2014. A knowledge-based ingredient formulation system for chemical product development in the personal care industry. Comput. Chem. Eng. 65, 40–53. https://doi.org/10.1016/j.compchemeng.2014.03.004

Smith, B. V., Ierapepritou, M.G., 2010. Integrative chemical product design strategies: Reflecting industry trends and challenges. Comput. Chem. Eng. 34, 857–865. https://doi.org/10.1016/j.compchemeng.2010.02.039

Williams, H.P., 2013. Model building in mathematical programming, 5th ed. John Wiley & Sons, Inc., Chennai.

536301 Investigating Fluid-Particle Interactions in Expanded Beds Using CFD-DEM

Abstract:

The purification of high value products such proteins produced via fermentation typically involves a cascade of unit operations such as membrane filtration and chromatography columns. Expanded bed adsorption (EBA) creates an opportunity for process intensification by combining cell removal with a product capture and initial purification step. By mildly fluidizing a bed of adsorbent particles, the inter-particle void space is increased, which allows cells and other particulate matter to move through the bed in a relatively unhindered manner. The bed is stabilized through the use of a particle size and density distribution, which cause bigger/heavier particles to remain in the bottom while smaller/lighter particles reside higher up in the column. This results in a gradient in particle size, particle density and particle volume fraction along the height of the bed which reduces particle movement and thereby reducing unwanted back mixing in both phases.

To successfully scale up this technology to production scale, a more fundamental understanding of the mechanisms behind fluidization in EBA columns is required. Research into expanded beds using experimental techniques yields different conclusions based on whether the solid or liquid phase is examined (1, 2). Given the nature of expanded bed systems (dense, opaque multiphase flows), non-intrusive measurements of the fluid-particle interaction at the particle scale are highly challenging.

As an alternative to experimental measurements, computational fluid dynamics (CFD) can be combined with a discrete element model (DEM) for the particle phase. This work will present the results of CFD-DEM simulations on lab scale columns and compare them with experimental data. It will be demonstrated how this approach can be used to link the mesoscopic behaviour of the particles to macroscopic phenomena such as bed expansion, fluid phase dispersion and velocity. Based on the insights obtained from the simulations, resin properties such as density and size distribution can optimized to ensure bed stability and separation performance. The overall goal is to understand, optimize and scale EBA technology which then can be used for various business applications, like bio-pharmaceuticals, natural products and enzymes. References

1. Souquet J, Liu HY, Fan XF, Theodosiou E, Thomas O. Exposing expanded beds to PEPT. Journal of Biotechnology. 2007;131(2):S191-S.
2. Lin DQ, Shi W, Tong HF, van de Sandt EJ, Boer P, Ferreira GN, et al. Evaluation and characterization of axial distribution in expanded bed: II. Liquid mixing and local effective axial dispersion. J Chromatogr A. 2015;1393:65-72.

522822 Mechanistic Study on C-C Coupling of Acetaldehyde on Partially Reduced $CeO_{2-X}(111)$

Abstract:

Selective C-C coupling catalyzed by solid catalysts is pertinent to the manufacture of fuel and chemical products from renewable or available carbon sources such as biomass and C1 compounds including methane and CO_2. A versatile way to accomplish C-C coupling between organic oxygenates is aldol addition/condensation. Here we use density functional theory calculations and microkinetic modeling to gain fundamental insights into how ceria, an earth-abundant solid material, catalyzes C-C coupling reactions at the molecular level, with acetaldehyde (AcH) as a representative carbonyl compound. We show how an enolate-based mechanism is capable of explaining the main features of temperature programmed desorption of AcH on $CeO_2(111)$[1] and how it suggests AcH might be coaxed into undergoing C-C coupling to form crotonaldehyde under UHV conditions, which is experimentally confirmed. We then extend the UHV mechanisms to account for finite pressure effects and compare the expanded version with one proposed by Mann et al.[2] Important features of our proposed reaction mechanisms[3,4] include oxygen vacancy-catalyzed enolization of AcH, surface-mediated intermolecular H transfer, and dynamic behavior of oxygen vacancies, which may be generally relevant to organic reactions on reducible oxides.

1. L. Chen, D.R. Mullins, *J. Phys. Chem. C* **2011**, 115, 3385.
2. K.P. Mann, Z. Wu, F.C. Calaza, S.H. Overbury, *ACS Catal.* **2014**, 4, 2437.
3. C. Calaza, Y. Xu, D.R. Mullins, S.H. Overbury, *J. Am. Chem. Soc.* **2012**, 134**,** 18034.
4. Zhao C, Y. Xu, *Top. Catal.* **2017**, 60, 446.

526972 Transition Metal Oxides As Catalysts in the Diels-Alder Reaction between Furan and Methyl Acrylate

Abstract:

We investigate catalytic effects of transition metal oxides such as HfO_2 and ZrO_2 on the Diels-Alder reaction of furan and methyl acrylate forming the oxanorbornene carboxylic acid methyl ester. Our theoretical and experimental studies confirm that transition metal oxides exhibit reactivity in the Diels-Alder reaction. Furthermore, we theoretically investigate the effects of surface hydroxylation on the activities for both Langmuir and Eley-Rideal mechanisms. Our calculations suggest that Eley-Rideal is the favorable mechanism for this reaction on all studied surfaces. Clean, partially hydroxylated and fully hydroxylated HfO_2 and ZrO_2 surfaces lower the activation barrier of the Diels-Alder reaction between furan and methyl acrylate to different degrees. While it turned out that clean oxide surfaces are weak catalysts, surprisingly partially hydroxylated oxide surfaces show higher activities than fully hydroxylated ones. Based on our extensive studies, surface hydroxylation can be used as a means to enhance catalytic effects of transition metal oxides on the Diels-Alder reaction of furan and methyl acrylate. This is quite interesting from application point of view because metal oxides tend to be hydrated. This study shows that there is no need for catalyst dehydration prior to use in a catalytic cycle. Calculated activation barriers are in very good agreement with those obtained from experiments. Although the obtained turnover frequencies are lower than Hf- and Zr-BEA zeolites, this finding paves the road for more applicable and scalable methodologies to derive powerful Diels-Alder intermediates.

531249 Design of Solvent Composition for Acid-Catalyzed Reactions of Biomass-Derived Oxygenates Using Molecular Simulation-Derived Observables

Abstract:

The aqueous-phase, acid-catalyzed conversion of lignocellulosic biomass is a promising method for the production of liquid fuels and high-value commodity chemicals from renewable feedstocks, but achieving sufficient product selectivity and yields remains an ongoing challenge. One simple strategy to improve reaction yields is to modify the solvent composition, which affects reaction rates, product selectivity and stability, and the economics of downstream separations. However, identifying an optimal solvent composition empirically by trial-and-error is cost-prohibitive and gives little insight into how the solvent environment will perform in new processes. We will describe our efforts to use classical atomistic molecular dynamics (MD) simulations to predict the effect of solvent composition on the conversion of biomass-derived model compounds. Leveraging the computational efficiency of MD simulations, we simulate a range of solvent mixtures containing water and polar aprotic cosolvents in varying mass fractions. Based on experimentally determined reaction rates, we show that as the water content of the solvent environment decreases, reactants with more hydrophilic groups have improved catalytic turnover rates for both hydrolysis and dehydration reactions. We explain this behavior by using MD simulations to quantify the size and properties of water-enriched local domains in the vicinity of the reactant in mixed-solvent environments. We further demonstrate that multiple simulation-derived observables can be combined into a predictive model that correlates with experimentally determined reaction rates for systems with varying solvent compositions. Finally, we demonstrate that analysis of the spatial distribution of solvent molecules can inform our understanding of reaction selectivity. These computational models are important for enabling the rational design of new liquid-phase biomass conversion processes by guiding the selection of specific cosolvents.

531409 Catalytic Hydrogenation of Carbon Monoxide to Formaldehyde in Functionalized Metal Organic Frameworks: An Investigation of Pathway and Uncertainty

Abstract:

Formaldehyde is a highly desirable platform chemical with an annual demand of over 30 million metric tons. The current most common method of generating formaldehyde industrially involves conversion of methanol to formaldehyde and suffers from low exergy efficiency.[1] The methanol used is typically produced by the oxidation of methane to syn-gas, followed by the reduction of CO to methanol via Fischer-Tropsh synthesis. In this work, we will present a direct pathway for converting CO to formaldehyde by direct hydrogenation in Metal Organic frameworks (MOFs) functionalized with Lewis pairs (LPs). MOFs are a class of porous materials having potential for both gas adsorption and catalysis. Functionalization of MOFs with catalytic moieties results in heterogeneous catalysts with active sites similar to those of molecular catalysts

Based on our previous studies, the one of the biggest advantage of using LPs as hydrogenation catalyst is the fact that LPs are good at heterolytic hydrogen dissociation. [2-4] In those pathways, the concerted transfer of a hydride and proton to the C and O atoms (respectively) of CO_2 had a much lower activation barrier than sequential addition of hydrogens to produce formic acid. In this work we test whether similar pathways can be used to directly reduce CO to formaldehyde with low barriers. The density functional theory (DFT) results presented here employed the GPAW code, with the BEEF-vdW functional [5] to describe the exchange correlation energy. We computed reaction pathways for H_2 dissociation, CO chemisorption, CO hydrogenation to formaldehyde and methanol. Using the BEEF ensemble, we also quantified the uncertainly associated with each DFT calculation due to the choice of exchange-correlation functional.

We have investigated the proposed pathway in gas phase, inside functionalized UiO-67, and on a freestanding fragment containing the LP ($N\text{-}BF_2$). We found that the poisoning of the LP by CO is reduced when the LPs were incorporated into the MOF. The rate-limiting step is the hydrogenation of CO, with a barrier of 1.30 ± 0.15 eV. We believe that conversion of CO to formaldehyde could be further enhanced in MOFs due to the increase in effective pressure inside confined pores. Additionally, this confinement effect also reduces translational degrees of freedom, which makes the free energy of the overall reaction more favorable.

1. Bahmanpour, A., Hoadley, A., Tanksale, A. Critical Review and Exergy Analysis of Formaldehyde Production Processes. *Reviews in Chemical Engineering*, 30(6), pp. 583-604. (2014).
2. Ye, J.; Johnson, J. K. *ACS Catal.* 5, 6219-6229 (2015)
3. Ye, J.; Johnson, J. K. *Sci. Technol.* 6, 8392-8405 (2016)
4. Ye, J.; Johnson, J. K. *ACS Catal.* 5, 2921-2928 (2015)
5. Wellendorff, J. et al. *Phys. Rev. B.* 85, 235149 (2012)

532307 Theoretical Insights into Catalytic Upgrading of Ethanol over 2D MFI Zeolite

Abstract:

The acid-catalyzed conversion of ethanol over zeolite has been considered as one of the promising routes for producing desirable liquid hydrocarbons. In particular, the reaction pathway near Brønsted acid site (BAS) proceeds via multiple mechanisms, including dehydration of ethanol to ethene with subsequent dimerization to butene or higher olefins. Changing the zeolite morphology from 3D framework to 2D lamella has been known to impact such Brønsted acid chemistry, ultimately affecting the product reactivity/activity of zeolite catalyst. As our first step to understanding the effects of zeolite morphology, we have examined the mechanistic steps of ethanol conversion at the interior and exterior of 2D MFI-based zeolite model. Using density functional theory (DFT) calculations, the relative binding energetic of reaction intermediates were computed at BAS. Our DFT calculations showed that the binding energies of intermediates are similar at the interior and exterior sites. The reaction pathway and free energy profile were further investigated by performing ab initio molecular dynamics (AIMD) simulations at experimental conditions ($T \sim 425$ °C). The enthalpic and entropic contribution to the free energy at the interior and exterior sites will be discussed in this work. We will show that the rate-determining steps and product selectivity towards higher hydrocarbons are significantly affected by the morphology of zeolite pores.

535167 Multicomponent Catalysis: Directing Reaction Pathways for Hydrodeoxygenation of Furfuryl Alcohol at Pd/TiO_2 Interfaces

Abstract:

Biomass has potential to generate liquid fuels after processing to reduce oxygen content. Catalytic hydrodeoxygenation(HDO) uses a hydrogen rich environment to deoxygenate, generating products with suitable chemical and thermal stabilities without any loss of carbon. Biomass-derived furfural is a fuel precursor and model compound for exploring HDO. Furfural HDO preferably leads to methyl furan (a potential fuel additive). Recently, core-shell catalysts where individual noble metal nanoparticles (e.g. Au, Ag, Pt, and Pd) are isolated in the core/yolkâ €"shell nanostructure of metal oxides (e.g. TiO_2, CeO_2, and ZrO_2) claim to bring unique collective and synergetic function in comparison with single-component materials [1]. Palladium nanoparticles encapsulated by porous TiO_2 showed high selectivity and activity towards HDO and minimal non-selective decarbonylation (DC), with the optimal catalyst having the smallest TiO_2 pore sizes [2]. Despite this encouraging performance, the mechanistic role of TiO_2 or the porous encapsulation that dictates the improved selectivity is unknown. TiO_2 can both accept a proton from H_2 to generate active sites and donate a proton to assist in cleaving the Câ''O bond of hydroxylated reactants[3]. Therefore, interfacial sites between metals and metal oxides have been proposed to play a vital role in assisting this complex reaction pathway with amplified selectivity to HDO and also hindering the side reactions like decarbonylation or aromatic ring hydrogenation.

We use density functional theory (DFT) to study the elementary surface reactions of furfuryl alcohol at a series of models for the Pd-TiO_2 interface. Interfacial models include Ti_2O_4 clusters and rutile TiO_2 nanowires over Pd(111) and a surrogate Helium pore model. The Helium pore model provides an inert interfacial site to probe the effect of binding orientations of adsorbates on its reaction pathways[4]. Elementary reaction energetics using the pore model suggest the hindering of decarbonylation in constrained pores, emphasizing that DC requires a large ensemble of sites and, hence, most favourably occurs in flat lying conformations [5]. HDO kinetics remain unaffected even with the change in binding orientations. DFT kinetics studies over Ti_2O_4 clusters imply a reduction in activation barrier for conversion of furfuryl alcohol to methyl furan through direct deoxygenation, pointing out the coupled effect of components across the interface.

This fundamental understanding of the elementary processes occurring at the interfaces will have significant implications in developing novel catalysts that combine unique functionalities across materials for improved selectivity and activity.

References:

1. Li, G. and Z. Tang, *Noble metal nanoparticle@metal oxide core/yolk-shell nanostructures as catalysts: recent progress and perspective.* Nanoscale, 2014. 6(8): p. 3995-4011.
2. Zhang, J., et al., *Directing Reaction Pathways through Controlled Reactant Binding at Pdâ€"TiO_2 Interfaces.* Angewandte Chemie, 2017. 129(23): p. 6694-6698.
3. Nelson, R.C., et al., *Experimental and Theoretical Insights into the Hydrogen-Efficient Direct Hydrodeoxygenation Mechanism of Phenol over Ru/TiO2.* ACS Catalysis, 2015. 5(11): p. 6509-6523.
4. Gomez-Gualdron, D.A., et al., *A modelling approach for MOF-encapsulated metal catalysts and application to n-butane oxidation.* Physical Chemistry Chemical Physics, 2015. 17(41): p. 27596-27608.
5. Pang, S.H., et al., *Effects of Thiol Modifiers on the Kinetics of Furfural Hydrogenation over Pd Catalysts*, in *Acs Catalysis*. 2014. p. 3123-3131.

537483 Homogeneous Catalysis of Ketene Production By Triethylphosphate

Abstract:

A mechanism of pericyclic reactions involving metaphosphoric acid has been developed for the dehydration of acetic acid to form ketene using an organophosphate catalyst. Ketene is an important and highly reactive chemical precursor, including usage for the production of acetic anhydride. Dehydration of acetic acid with organophosphate catalysts has been observed and cited in industrial patents. The only elementary reaction mechanism of this sort that has been published was a study of catalysis by P_2O_3/P_2O_5 was published by Sebbar et al. (2016), addressing the reaction as a primarily radical reaction mechanism. This current work approaches the dehydration of acetic acid using triethylphosphate as the catalyst and considering only pericyclic reactions and demonstrates that the reaction proceeds through formation of the reactive intermediate metaphosphoric acid (HPO_3).

Stable intermediate species and transition-state structures included in the proposed mechanism were simulated using computational quantum chemistry software (Gaussian 09). Initial structure exploration and optimization were carried out using DFT methods at a B3LYP/6-31G(d,p) level of theory. Refined structures and their thermochemical properties were calculated with the composite method CBS-QB3. Hindered-rotation anharmonicity was considered for hydroxyl and methyl rotations in the structures using rotational energy profiles calculated at a B3LYP/6-31G++(d,p) level of theory. For electronically barrierless reactions, the transition state was calculated as the highest-free-energy structure in the reaction path, according to variational transition-state theory. Variational transition states were initially found using a B3LYP/6-31G++(d,p) level of theory and then refined using CBS-QB3. Rates of reaction were calculated using the Mesmer master equation code. Flow-reactor predictions using the resulting mechanisms and Chemkin Pro were compared against experimental flow-reactor yields as determined by molecular-beam mass-spectrometry. Potential pathways were evaluated by comparing reaction kinetics at the high-pressure limit against what was observed experimentally.

Eight possible pathways were considered. Only one pathway had rates fast enough to explain the experimental yields, the pathway involving metaphosphoric acid as a reactive intermediate. This mechanism begins with formation of phosphoric acid from triethylphosphate by successive eliminations of ethylene. Phosphoric acid then goes through a catalytic cycle in which it dehydrates to form metaphosphoric acid, associates with acetic acid to form acetylphosphate, and finally dissociates to yield ketene and a reformed phosphoric acid molecule. Predicted yields agreed with experimental yields within an order of magnitude without requiring any parameter fitting.

The authors gratefully acknowledge funding of this research by the Eastman Chemical Company and computational resources provided by the Extreme Science and Engineering Discovery Environment (XSEDE), which is supported by National Science Foundation grant number ACI-1548562.

533740 Connecting Experimental Conditions with Chemical Structure

Abstract:

Reaction databases provide a great deal of useful information to assist planning of experiments, but do not provide any interpretation or chemical concepts to accompany this information. In this talk, reactions are labeled with experimental conditions, and network analysis will show that consistencies within clusters of data points can be leveraged to interpret underlying concepts that organize the information. In particular, this analysis will show why particular choices of solvent are effective in enabling specific organic reactions (Friedel-Crafts, aldol addition, Claisen condensation, Diels-Alder, and Wittig), including variations in solvent choice within each reaction class. Our strategy will automatically craft succinct sentences or rules of thumb which describe the observed solvent selection rules. An example of network analysis is shown in Figure 1, where a Friedel-Crafts reaction breaks into clusters which depend on choice of catalyst, substrate, and stability of the reaction for/against protic solvent. These observations, typically belonging to the intuition of a chemist, are derived directly from the network.

Figure 1. Friedel-Crafts reactions network.

530881 Symbolic Regression of Alpha Functions for Cubic Equations of State

Abstract:

Since Van der Waal first proposed a formulation for a thermodynamic cubic equation of state, modifications to the thermodynamic cubic equations of state have focused on the alpha function of the attraction term to model the temperature dependence of attraction for real fluids [1]. The alpha function is dependent on reduced temperature and density and acentric factor. Modifications to Peng Robinson and Redlich-Kwong forms of the cubic equation have focused on the functional form of the alpha function fit prior to fitting the attraction and repulsion values for the cubic equation. Traditionally, cubic equations are derived from the critical point which leads to errors in liquid density and in the critical region. Alternatively, when fit to chemical-specific liquid density and vapor pressure data, the equations fail to represent the critical constraints [2].

To improve the development of cubic equations and their regression, we focus on a data-driven constrained symbolic regression to simultaneously determine the model form of the alpha function and fit parameters of the cubic equation. Symbolic regression learns both the model structure and parameters to model a data set, unlike traditional regression that limits the scope of the regression to a fixed functional form [4]. The regression only requires the specification of a set of operators and operands (+, -, *,÷, exp(·), log(·), $(\cdot)^2$, $(\cdot)^3$, $\sqrt{\cdot}$, etc.) to flexibly develop new functional forms that accurately represent the data. Symbolic regression has typically been performed with genetic programming [4], but recent developments in applying global deterministic approach have shown improved fitting metrics, such as sum of squared error and other information criteria [3]. We use this approach to apply symbolic regression to pure fluids to determine new alpha function forms for cubic equations of state and compare their accuracy to the data, specifically in liquid density and supercritical regions, and adherence to theoretical thermodynamic properties with other alpha modifications of cubic equations of state.

Reference cited:

[1] Valderrama, J. O. The State of the Cubic Equations of State. *Ind. Eng. Chem. Res.* 2003, 42, 1603-1618.

[2] Twu, C.H. & Sim, W.D. & Tassone, V. (2002). Getting a Handle on Advanced Cubic Equations of State. Chemical Engineering Progress. 98. 58-65.

[3] Cozad, A. & Sahinidis, N.V. (2018). A global MINLP approach to symbolic regression. Mathematical Programming, to appear.

[4] J. R. Koza. Genetic Programming: On the Programming of Computers by Means of Natural Selection. MIT Press, 1992.

538152 Predicting Hydrogen Storage in Half-a-Million MOFs Via Machine Learning

Abstract:

Metal-organic frameworks (MOFs) are promising solid-state adsorbents thanks to their high gravimetric capacities. However, realizing a high volumetric H_2 adsorption capacity, balanced with a high gravimetric density, is one of the main barriers of the successful application of MOFs as solid-state adsorbents. A large database of half-a-million MOFs consisting of real and hypothetical compounds was compiled and screened using semi-empirical and atomistic (grand canonical Monte Carlo) techniques. Several machine learning (ML) algorithms were benchmarked for their ability to predict hydrogen storage in MOFs at multiple conditions. The top performing algorithm, extremely randomized trees, was applied to rapidly identify MOFs with high usable H_2 storage capacities across the entire database. A combinatorial approach was then used to understand the importance of crystallographic properties and training set size. This approach identifies the number and combination of crystallographic features needed to achieve the most accurate predictions. Finally, we assess multilinear regression models for their ability to out-perform the well-known Chahine rule for predicting H_2 uptake.

537160 Inverse Design of Nanoporous Adsorbents for Gas Separation Applications

Abstract:

Significant research efforts have been undertaken for *in silico* screening of nano-porous materials in applications such as methane storage, carbon capture[1-2]. Both theoretical insights and data-driven approaches have been used to develop structure property relationships for properties such as gas storage, selectivity, heat of adsorption[3-6]. Balancing multiple material property tradeoffs is important for both commercial and novel applications. Optimizing a completely algebraic formulation of a process can provide optimal material property targets. While these may not correspond directly to an existing material, possible structures can be predicted inversely by exploiting similarity[7] between structures and properties of adsorbent materials.

In this work, the optimal adsorbent properties (e.g. adsorption isotherms, heat of adsorption etc.) are determined by optimizing a completely discretized non-linear programming model of a pressure swing adsorption process for a desired objective such as cost or productivity subject to process constraints such as purity and recovery. The bounds on the material properties are set based on the isotherm data obtained from Grand Canonical Monte Carlo (GCMC) simulations of structures in the IZA-SC database. This approach is advantageous as all the material alternatives do not have to be individually considered. Moreover, the optimal material properties can be recalculated for changes in process and feed conditions with relative ease. For the reverse estimation of structural properties, a surrogate model based approach[8] is employed. It relates the differences in structural properties such as channel diameter, pore size, accessible surface area and volume between pairs of pure silica zeolites to the difference in their adsorption properties. Based on this, the appropriate structural features corresponding to the target material properties are then obtained. The salient feature of this approach is to design materials for optimal operation of processes such as CO_2 capture, methane storage.

References:

(1) Colon, Y. J.; Snurr, R. Q. High-Throughput Computational Screening of Metal-Organic Frameworks. *Chem. Soc. Rev.*, **2014**, 43 (16), 5735–5749.

(2) Lin, L.-C.; Berger, A. H.; Martin, R. L.; Kim, J.; Swisher, J. A.; Jariwala, K.; Rycroft, C. H.; Bhown, A. S.; Deem, M. W.; Haranczyk, M. and Smit, B. In silico Screening of Carbon-Capture Materials. *Nat. Mater.*, **2012**, 11 (7), 633–641.

(3) Wilmer, C. E.; Farha, O. K.; Bae, Y-S.; Hupp, J. T. and Snurr, R. Q.; Structure–property Relationships of Porous Materials for Carbon Dioxide Separation and Capture, *Energy Environ. Sci.*, **2012**, 5, 9849-9856.

(4) Garcia, E. J.; Perez-Pellitero, J.; Jallut, C. and Pirngruber, G. D. How to Optimize the Electrostatic interaction between a Solid Adsorbent and CO_2. *J. Phys. Chem. C*, **2014**, 118, 9458-9467.

(5) Bai, P.; Jeon, M. Y.; Ren, L.; Knight, C.; Deem, M. W.; Tsapatsis, M. and Siepmann, J. I. Discovery of Optimal Zeolites for Challenging Separations and Chemical Transformations using Predictive Materials Modeling. *Nat. Commun.*, **2015,** 6, 5912.

(6) Fernandez, M.; Boyd, P.G.; Daff, T. D.; Aghaji, M. Z. and Woo, T. K. Rapid and Accurate Machine Learning Recognition of High Performing Metal-Organic Frameworks for CO_2 Capture. *J. Phys. Chem. Lett.* **2014**, 5, 3056-3060.

(7) Martin, R. L.; Willems, T. F.; Lin, L.-C.; Kim, J.; Swisher, J. A.; Smit, B.; Haranczyk, M. Similarity-Driven Discovery of Zeolite Materials for Adsorption-Based Separations. *ChemPhysChem,* **2012**, 13, 3595–3597.

(8) Bajaj, I.; Iyer, S. S. and Hasan, M. M. F. A Trust Region-based Two Phase Algorithm for Constrained Black-box and Grey-box Optimization with Infeasible Initial Point. *Comput. Chem. Eng.*, **2017**, doi: 10.1016/j.compchemeng.2017.12.011.

530836 Classifying Antimicrobial and Multifunctional Peptides with Machine Learning

Abstract:

Antimicrobial and antifouling peptides are highly desirable biomolecules, but they are relatively rare among all known peptides. It is not intuitively apparent from an amino acid sequence whether a given peptide will exhibit these properties. Thus, it is of interest to develop a data-driven quantitative structure-activity (QSAR) model to predict potentially antimicrobial and antifouling peptides. Machine learning techniques are well-suited to such a task, and have the advantage of a robust, established pool of algorithms for learning. Here, we exhibit Bayesian network QSAR models that allow us to accurately predict whether an input peptide may exhibit antimicrobial properties, given its sequence.

529531 A Systematic Procedure for Designing Training Data for Molecular Property Prediction

Abstract:

Organic material design requires thoroughly exploring chemical compound space to gain desired property information. The astronomical size of molecule space, however, makes it impossible to use experiments of quantum chemistry to evaluate every molecule in the space; consequently, data-driven semi-empirical models are required to calculate molecular properties rapidly. To this end, the focus of this talk is two-fold. First, we show that the group contributions approach can be generalized using a combination of Cheminformatics-based path fingerprints and sparse modeling techniques to derive an ab initio data-driven model to accurately estimate heats of atomization of small organic molecules, reaches the mean absolute error of 1.59 kcal/mol and 2.61 kcal/mol for QM7 and QM9 dataset excluding molecules with fused rings. Further, we show that modern experimental design tools and cheminformatics-based subset selection techniques can be combined to systematically minimize the amount of data needed to train a model. We specifically show that, given a set of molecules for which data-driven models are sought, a carefully chosen subset of even 2- 10% of original dataset is sufficient to train a model that is as accurate as one that is trained on >80% of the data.

528153 Designing Proteins with Enhanced Antifreeze Activity Using SimulatedÂ Directed Evolution

Abstract:

Found in a variety of organisms, from cold water fish to arctic beetles, antifreeze proteins (AFPs) are a diverse class of proteins that depress the non-equilibrium freezing point of water. AFPs have been of scientific interest for over 30 years due to their role in organism survival at low temperatures, and have potential applications in cryopreservation, cryosurgery, clathrate formation inhibition, and specialty coatings. Despite a large body of experimental and theoretical literature accumulated on AFPs, there exists no widely applicable method for accurately predicting AFP activity - making intelligent design of new antifreeze materials challenging. To remedy this, we perform molecular dynamics simulation with a collection of different types of AFPs (ranging from ~30 to 300 residues in length) in liquid water just below the freezing point. By analyzing the dynamics of hydrogen bonds in water near the protein in conjunction with the geometry of the protein structure, we establish a general and quantitatively accurate correlation between simulation and experimentally measured antifreeze activity. Implementing this correlation in a directed evolution algorithm, we identify several key mutations that exhibit more than five time the predicted activity than the wild type protein.

538791 Transition State Geometry Prediction Using Neural Embeddings of Transition State Graphs

Abstract:

Transition State Theory (TST) with modern quantum chemistry methods allow reaction rates coefficients to be calculated with high accuracy, but finding the transition state is computationally expensive and difficult to automate. The key step in TST is to determine the transition state (TS) geometry, which is mostly determined by reaction family, reaction center, and reactant geometries. Inspired by neural word embeddings techniques, we introduce rxn2vec, which is an unsupervised machine learning approach to create vector representations of transition states. We build upon a neural embedding framework named graph2vec [1] to learn data-driven distributed representations of arbitrarily sized graphs. In this case, fully optimized TS geometries generated by our automated TST calculator (AutoTST) [2] are converted to undirected graphs using connectivity matrices. These training data then provide the framework to predict key TS reaction center geometries. We demonstrate our tool on a set of reactions, predict key distances from these reaction, and implement AutoTST to perform partial geometry optimizations to converge on a finalized geometry. So far this tool can be used to predict the reaction centers of three reaction families, namely hydrogen abstraction, intramolecular hydrogen migration, and radical addition to multiple bond.

[1] graph2vec: Learning Distributed Representations of Graphs
A. Narayanan, M. Chandramohan, R. Venkatesan, L. Chen, Y. Liu, S. Jaiswal (2017)
arXiv:1707.05005 [cs.AI]

[2] Automated Transition State Theory Calculations for High-Throughput Kinetics
Pierre L. Bhoorasingh, Belinda L. Slakman, Fariba Seyedzadeh Khanshan, Jason Y. Cain, and Richard H. West. J. Phys. Chem. A.121 (37), 6896-6904, (2017) https://doi.org/10.1021/acs.jpca.7b07361

525270 Predicting Virial Coefficients and Alchemical Transformations By Extrapolating Mayer-Sampling Monte Carlo Simulations

Abstract:

A statistical mechanical extrapolation method is presented which allows a single Mayer-sampling Monte Carlo simulation to predict virial coefficients over a large range of both temperatures and model parameter values (i.e., alchemical transformation) with minimal increase in computational cost [1]. We will show that a simulation of the SPC/E (extended simple point charge) water model at one state point can quantitatively predict the second virial coefficient continuously across three orders of magnitude in temperature change, concomitantly changing by four orders of magnitude in value, with less than a 2% deviation. The same simulation was also used to simultaneously predict the second virial coefficient of a modified SPC/E model with site charges scaled by a constant factor, from an increase of 40% down to zero charge. This method is also shown to perform well for the third virial coefficient and the exponential model parameter for a Lennard-Jones fluid. Example code is provided at https://github.com/usnistgov/mayer-extrapolation.

[1] H.W. Hatch, et al., J. Chem. Phys., 147, 231102 (2017). https://doi.org/10.1063/1.5016165

537323 First Principles Monte Carlo Simulations of Adsorption and Reaction Equilibria

Abstract:

This talk will highlight recent developments enabling first principles Monte Carlo (FPMC) simulations for which the potential energy is calculated on-the-fly using Kohn-Sham density functional theory. Applications of FPMC to the prediction of (a) adsorption isotherms for gas molecules in metal-organic frameworks with under-coordinated metal nodes and (b) reaction equilibria in cation-exchanged zeolites. Emphasis will be given to simulation methodologies and microscopic-level origins of the observed thermodynamic behavior.

532680 Molecular Exchange Monte Carlo: A Generalized Method for Identity Exchanges in Grand Canonical Monte Carlo Simulations

Abstract:

A generalized identity exchange algorithm is presented for Monte Carlo simulations in the grand canonical ensemble. The algorithm, referred to as Molecular Exchange Monte Carlo (MEMC), may be applied to multicomponent systems of arbitrary molecular topology, and provides significant enhancements in the sampling of phase space over a wide range of compositions. Three different approaches are presented for the insertion of large molecules, and the pros and cons of each method are discussed. The performance of the algorithms is highlighted through grand canonical Monte Carlo histogram-reweighting simulations of binary mixtures. In simulations of the binary mixture of perfluorobutane+butane, the acceptance rate for MEMC moves is up to 258 times that of configurational-bias insertions and deletions, producing two orders of magnitude improvements in computational efficiencies. Calculations performed for mixtures of methane with various n-alkanes show increases in acceptance rates for molecule exchanges of 2-70 times that of single molecule insertions, with corresponding order of magnitude improvement in computational efficiency. Comparisons to other methods, such continuous fractional component Monte Carlo are made.

532273 Computational Cluster-Integral Methods for Solutions

Abstract:

Cluster-integral ideas underlie many theoretical methods in statistical mechanics. The most widely recognized and widely used approach is in the form of the virial equation of state. This framework identifies coefficients of a density expansion of the pressure (and other properties) in terms of integrals of multimolecular interactions in a vacuum. The pairwise-additive intermolecular potential forms frequently encountered in molecular simulations are often used in this context. A corresponding framework is available for properties of solutions. Here the expansion is (typically) for the osmotic pressure, as a series in solute concentration. The coefficients again can be given as cluster integrals, but now the interactions are among solutes as mediated by their interactions with the solvent. Typically a pairwise-additive approximation for solute-solute interactions is not as effective in such applications. Numerical evaluation of the indicated cluster integrals is not as easily accomplished, and require methods different from those that have been developed for gas-phase virial coefficients. In this paper we present and examine novel and very effective methods, and demonstrate them for several model systems. We consider in particular the performance of the methods, the nature of the effective multibody interactions, and the effectiveness of the overall framework for describing solution properties.

537029 Nucleus-Size Pinning for Determination of Nucleation Free-Energy Barriers and Nucleus Geometry

Abstract:

Classical Nucleation Theory (CNT) has recently been used in conjunction with a seeding approach to simulate nucleation phenomena at small-to-moderate supersaturation conditions when large free-energy barriers ensue. In this study, the conventional seeding approach [J.R. Espinosa, C. Vega, C. Valeriani, and E. Sanz, J. Chem. Phys. 144, 034501 (2016)] is improved by a novel, more robust method to estimate nucleation barriers. Inspired by the interfacial pinning approach [U. R. Pedersen, J. Chem. Phys. 139, 104102 (2013)] used before to determine conditions where two phases coexist, the seed of the incipient phase is pinned to a preselected size to iteratively drive the system toward the conditions where the seed becomes a critical nucleus. We call this technique nucleus-size pinning (NSP). The proposed technique is first validated by estimating the critical nucleation conditions for the disorder-to-order transition in hard spheres, and then applied to simulate and characterize the highly non-trivial (prolate) morphology of the critical crystal nucleus in hard gyrobifastigia. A generalization of CNT is used to account for nucleus asphericity and predict nucleation free-energy barriers for gyrobifastigia. These predictions of nuclei shape and barriers are validated by independent umbrella sampling calculations. With another application of CNT, we are able to determine interfacial tensions for these systems within a single simulation, which match those estimated by standard techniques. Further, we discuss applications of NSP for solid phase nucleation in other systems.

532189 A Free-Energy Diabat Approach to Polymorph Stability

Abstract:

Many molecules can crystallize into several different crystal structures (polymorphs), each with different solubility and growth/dissolution kinetics. There are several methods to discover and predict the free-energies of polymorphs. All the methods, however, rely on quasi-harmonic approximations or on expensive multi-step simulations. We propose a framework that combines the idea of Bennettâ€™s diabat interpolation method and energy gaps from lattice-switch Monte Carlo techniques previously used only in electron transfer literature. We illustrate that only two unbiased simulations are required for the Helmholtz free-energy difference between FCC and BCC polymorphs of Gaussian core solid. The method is generalized for the Gibbs free-energy difference in the isothermal-isobaric ensemble. The relevant Zwanzig-Bennett relation in the NPT ensemble is applicable for the diabats of triclinic polymorphs. The simple interpolation idea is not very accurate for anharmonic diabats, however the two unbiased simulations, do provide preliminary bias potentials for umbrella sampling. The flattened landscape can then be efficiently sampled. We illustrate the method for zirconium (BCC and HCP phases) and benzene molecular crystals (form I and form III).

538144 Free Energy Landscape with Experiment Directed Simulations and Enhanced Sampling

Abstract:

Free Energy Surface calculations for protein folding is a daunting challenge in Molecular Dynamics (MD) without any enhanced sampling or biasing techniques. Due to large energy barriers between local and global minima, molecules may be trapped in local minima during regular MD, which is called a sampling issue. To overcome this issue, enhanced sampling techniques, such as, parallel tempering well-tempered ensemble (PT-WTE) can be used in protein simulations. PT-WTE is a multiple replica simulation running at different temperatures and swapping conformations between replicas by forcing the simulation to visit states that have not been explored yet. Simulations can be further modified to match some experimental bias. To improve the agreement between simulations and experiments, simulations can be biased with Experiment Directed Simulations (EDS). EDS is a single replica biasing technique that converges the ensemble averages of collective variables in a simulation and experiments by using maximum entropy method. As a model system for proteins, GYG peptideâ€™s free energy surface was predicted by biasing the simulations with EDS in order to match experimental chemical shifts and by applying PT-WTE. EDS has shown compelling improvement in predicting free energy surface and efficiently sampled global minima with enhanced sampling.

538634 No More Histograms: Variational and Bayesian Approaches to Estimating Potentials of Mean Force

Abstract:

Potentials of mean force, or free energies along a selected set of collective variables, are ubiquitous in molecular simulation. Applications include determining the rate limiting step of a reaction, understanding the behavior of collective interactions such as hydrophobicity, and elucidating them mechanism of transport along molecular pores. Most commonly, the potential of mean force is estimated using a range of different histogramming techniques, most often a type of multiple histogram reweighting technique such as WHAM.

However, the process of histogramming obscures two important points; first, the observed distribution of observations along the desired collective variable or variables is not a histogram, but a series of delta functions, and approximating it as a histogram results in a loss of information. Second, the potential of mean force is (for molecular problems) a continuous function, and representing this function as a histogram is also an approximation.

In this study, we examine how to properly relate the observed empirical distribution to true, infinite sampling probability distribution as a function of collective variables, and hence determine the potential of mean force. We first show how histogramming is essentially a kernel density approximation with a "top hat" kernel, and show how we can generalize this process to other kernels that might be more useful for a given problem.

More powerfully, we present a variational approach, in which finding the continuous probability density function along a collective variable reduces to minimizing the Kullback-Leibler divergence between a trial function and the empirical distribution, as well as a fully Bayesian approach to sampling in the space of trial functions. This Bayesian approach additionally provides several powerful tools, such as determining which trial functions may be most appropriate for the problem and which trial functions are under or overfit. We demonstrate the application of these methods in 1, 2, and 3D model problems.

538781 Approaches to Finding Optimal Pathways and Flux Along Them in Multidimensional Free-Energy Hypersurfaces

Abstract:

Molecular-scale investigations of complex (bio)physical systems necessitate rationalizing reaction mechanisms by obtaining numerical estimates of free-energy differences between all conformational states of interest. To account for many relevant degrees of freedom in flexible molecules, it is often desired to construct free-energy hypersurfaces in a large yet finite number of variables because the knowledge of these hypersurfaces is useful in obtaining information on transition pathways and reactive fluxes along them. Contrary to several path-based simulation methods in which an initial guess of a parameterized path is iteratively optimized to obtain minimal free-energy pathways, in this work we explore initial applications of new simulation methods that allow searches within multidimensional free-energy surfaces by propagating solutions normal to a seed point to obtain the first exit points and hence pathways of minimal resistance. We will present results from the applications of these methods to model-potentials and analytically reconstructed free-energy hypersurfaces.

533192 Examination of Brønsted and Lewis Acid-Catalyzed Alkane Reactions in MFI Zeolites Using a Hybrid QM/MM Scheme

Abstract:

MFI Zeolites are particularly well-suited for the valorization of light alkanes as aromatic compounds, due to their shape selectivity to Benzene, Toluene and Xylene (BTX) products and resistance to coke formation [1]. Both H-MFI and metal-exchanged H-MFI zeolites, such as Ga/H-MFI, Zn/H-MFI, etc. are capable of alkane aromatization. However, Brønsted Acid sites present in these zeolite catalysts are known to be less effective for catalyzing alkane dehydrogenation chemistry [2, 3]. Using QM/MM computations which accurately capture molecular adsorption in the MFI Framework, we predict the role of Lewis Acid sites in zeolite-catalyzed alkane transformations; with the help of both experimental and computational efforts, we demonstrate how Lewis Acidic Gallium Hydrides provide effective electronic stabilization to rate determining C-H Cleavage bottlenecks. We additionally investigate alkane reactions and zeolites which were not investigated in our earlier work [4].

[1] J. Weitkamp, The Application of Zeolites in Catalysis. *Basic Princ. Appl. Catal.*, 161–212 (2004).

[2] T. F. Narbeshuber, H. Vinek, J. A. Lercher, Monomolecular Conversion of Light Alkanes over H-ZSM-5. *J. Catal.* **157**, 388–395 (1995).

[3] A. Bhan, W. Nicholas Delgass, Propane Aromatization over HZSM-5 and Ga/HZSM-5 Catalysts. *Catal. Rev.* **50**, 19–151 (2008).

[4] E. Mansoor, M. Head-Gordon, A.T. Bell, Computational Modeling of the Nature and Role of Ga Species for Light Alkane Dehydrogenation in Ga/H-MFI, Submitted. (2018).

530376 First Principles Modeling of Extended Solvent Structures in Defected Microporous Materials and Their Influence on the Kinetics of Lewis Acid Site Speciation

Abstract:

Microporous zeolites have found wide success in the petroleum industry as selective catalysts for refining fuels. In addition to BrÃ¸nsted acid zeolites, which mediate chemical reactions through protons, Lewis acid zeolites have been synthesized with framework-substituted metal cations such as Sn substituted zeolite Beta (Sn-Beta). Sn-Beta has been found to selectively catalyze oxygenate chemistries such as glucose isomerization,[1] in which Sn atoms within the framework generate catalytically active sites. Precise determination of the structure and speciation of the active Sn site is, however, complicated by the propensity for Sn atoms to adopt one of many possible â€œopenâ€ configurations that are characterized by the presence of a hydroxy ligand and framework silanol. Further, in addition to imparting catalytic active sites, Sn substitution changes the microporous voids in Beta from hydrophobic to hydrophilic in character by addition of polar binding sites for solvent. This transition introduces an additional parameter for catalyst design based on solvent interactions with reactive intermediates. While the structure of water has been described previously in defect-free zeolites[2,3], we provide the first detailed analysis using *ab-initio*
structure and thermodynamics at defects relevant to Lewis acid zeolites. The size and degree of hydrogen bonding of water structures at different chemical potentials and different defect types is considered and used to construct rational solvent models.

Experimental IR spectroscopy is used to observe the changes in hydrogen bond networks in low-defect and high-defect materials compared to those calculated with AIMD. We find that defect-free materials have low water uptakes at ambient conditions, but Sn sites nucleate water clusters. Silanol nests, formed through silicon vacancies introduced during synthesis, generate large extended water networks. The degree of clustering and transience of hydrogen bonding is analyzed using a graph theory approach to define membership rules for water clusters and extended networks. These results are then used to study the dissociation of water to form open Sn sites and study the role of liquid water on the relative populations of open and closed sites. Control of the relative populations of open and closed Sn sites, as well as the structure of the surrounding water network, has direct implications on the selectivity and rate of reactions catalyzed in microporous materials.

References:

(1) Bermejo-Deval, R., et al. *PNAS*. 109, 9727-9732 (2012).

(2) Demontis, P., et al. *Phys. Chem. B*. 107, 4426-4436 (2003).

(3) Zhou, T., et al. *Phys. Chem. C*. 121, 22015-22024 (2017).

530448 Computational Screening of Metal-Organic Frameworks for Direct Methane to Methanol Conversion

Abstract:

In the United Sates, the demand for natural gas as a chemical feedstock is currently high and predicted to continue to rise in the future.[1] However, the industrial method for transforming methane, the main constituent of natural gas, to valuable chemicals such as methanol is inefficient and can only be carried out on a large scale. In contrast, biomimetic catalyst design strategies employing supported iron or copper oxide nanoclusters allow for direct conversion of methane to methanol under mild temperature and pressure conditions.[2] Recent experimental and computational work has investigated a variety of approaches for generating iron and copper oxide nanoclusters, for example on zeolites,[3] metal-organic framework (MOF) nodes,[4] and porphyrin linkers in MOFs.[5] Here, we explore a new possibility where copper oxide nodes in MOFs may be used directly as the active sites for the reaction of interest. By means of Density Functional Theory calculations, MOFs with relevant copper node motifs were evaluated for important reaction steps such as methane activation and methanol desorption. Our goal is to identify potential candidates for further experimental verification and derive key activity descriptors that can provide insights into structure-property relations of active catalysts for direct methane to methanol conversion.

(1) Energy Information Administration. *Annual Energy Review*; U.S. Department of Energy, **2018**.
(2) Groothaert, M. H.; Smeets, P. J.; Sels, P. F.; Jacobs, P. A.; Schoonheydt, R. A. *J. Am. Chem. Soc.* **2005**, *127*, 1304-1395.
(3) Mahyuddin, M. H.; Staykov, A; Shiota, Y; Miyanishi, M; Yoshizawa, K. *ACS Catal.* **2017**, *7*, 3741â€"3751.
(4) Ikuno, T; Zheng, J; Vjunov, A; Sanchez-Sanchez, M; Ortuno, M. A.; Pahls, D.; Fulton, J. L.; Camaioni, D. M.; Li, Z; Ray, D.; Mehdi, B. L.; Browning, N. D.; Farha, O. K.; Hupp, J. T.; Cramer, C. J.; Gagliardi, L.; Lercher, J. A. *J. Am. Chem. Soc.* **2017**, *139*, 10294â€"10301.
(5) Doan, H. A.; Li, Z.; Farha, O. K.; Hupp, J. T.; Snurr, R. Q. *Catal. Today* **2018**. Advance Online Publication. https://doi.org/10.1016/j.cattod.2018.03.063.

531218 Computational Prediction of the Structure and Catalytic Properties of Copper Zirconium Oxide

Abstract:

Global warming presents an existential threat to humanity, requiring the development of methods which can reduce atmospheric levels of greenhouse gases such as CO_2. Moreover, conversion of CO_2 to other, useful products such as H_3COH is economically desirable. Our prior work[1,2] has demonstrated the ability of bimetallic CuZr nanoparticles to adsorb and activate CO_2, even when oxidized to $CuZrO_3$ (due to the oxophilicity of surface Zr sites). Further investigation of these oxidized NPs has been stymied by an absence of the oxideâ€™s detailed structure from literature. Using a genetic algorithm geared towards structure prediction (USPEX) in conjunction with planewave Density-Functional Theory, we present a prediction of the $CuZrO_3$ structure, which is to the best of our knowledge the first attempt to structurally characterize this mixed-oxide material in literature. Further, we investigate the adsorption of CO_2 to the predicted structure and provide a detailed comparison to experimental CO_2 adsorption studies.

1. Austin, N.; Ye, J.; Mpourmpakis, G. CO_2 Activation on Cu-Based Zr-Decorated Nanoparticles. *Cat. Sci. Tech.* **2017**, 7, 2245-2251.
2. Dean, J.; Yang, Y.; Austin, N.; Veser, G.; Mpourmpakis, G. Design of Copper-Based Bimetallic Nanoparticles for Carbon Dioxide Adsorption and Activation. *ChemSusChem* **2018**, 11, 1169-1178.

533678 Elucidating the Role of Oxygen Coverage in CO_2 Reduction on Mo_2C

Abstract:

Molybdenum carbide (Mo_2C) is a promising and low-cost catalyst for the conversion of CO_2 to CO. However, the underpinning mechanism occurring on the catalyst surface and the surface characteristics of Mo_2C catalyst under the reaction conditions are still elusive. In this study, we employ first principles calculations to understand the CO_2 dissociation mechanism on Î²-Mo_2C (001) under different oxygen coverage on the catalyst surface (oxycarbide).[1, 2] Our results demonstrate that with increasing oxygen coverage, there is an electronic modification on the catalyst surface (e.g. d-band shift on Mo atoms), that in turn, tunes the interactions of the adsorbates and the CO_2 dissociation barriers. Interestingly, we reveal linear relationships between the oxygen coverage and electronic modification with the reactivity of the catalyst. We show that CO_2 can adsorb and dissociate on the oxygen covered Mo_2C surface, even in the presence of surface oxygen up to 0.5 monolayer (ML). Our results rationalize a series of experimental observations.

References:

1. M. Dixit, X. Peng, M. D. Porosoff, H. D. Willauer and G. Mpourmpakis, *Cat. Sci. Tech.*, 2017, **7**, 5521-5529.
2. M. D. Porosoff, J. W. Baldwin, X. Peng, G. Mpourmpakis and H. D. Willauer, *ChemSusChem*, 2017, **10**, 2408-2415.

537843 Balancing Reactivity and Stability in Metal Nanoparticle and Alumina Support Systems Via Redox Reactions: A Multiscale Computational and Experimental Approach

Abstract:

Pt nanoparticles (NPs) supported on γ-Al_2O_3 (Pt/γ-Al_2O_3) serve in catalytic applications ranging from methanol reformation to CO oxidation, benefiting from the large surface area and strong metal-support interaction of γ-Al_2O_3 capable of stabilizing NPs. Additionally, Pt/γ-Al_2O_3 catalyst systems benefit from the high catalytic reactivity of small, nanoscale metal Pt NPs, the high surface-to-volume ratios of which enable many available sites for catalytic reactions and bonding to supports. However, under the reaction conditions – such as high (>600 °C) temperatures or highly oxidizing environments – frequently employed in industrial catalytic applications, Pt NPs are affected by issues such as poor long-term thermal stability and catalytic deactivation from Pt oxide formation. Solutions attempting to address these issues, such as the introduction of a reducing H_2 environment to remove Pt oxides, can change Pt NP morphology and reduce the effective surface area of the Pt/γ-Al_2O_3 interface, weakening the metal-support interaction and further destabilizing the catalyst system. In contrast, the addition of lower pressure oxidative pretreatments of O_2 or H_2O to Pt/γ-Al_2O_3 systems can further stabilize Pt/γ-Al_2O_3 via Pt-O and Pt-OH interfacial bonding. Thus, for Pt/γ-Al_2O_3 systems, catalysts can be rationally designed to be both stable and reactive by balancing the presence of reducing and oxidizing environments.

Though both O vacancies on oxide supports and metal-metal or metal-oxide bonding can stabilize metal-support catalyst systems, comparisons between experimental and computational Energy Electron Loss Spectroscopy (EELS) results show only metal-metal and metal-oxide bonding are found to be stable at the Pt/γ-Al_2O_3 interface. Therefore, computational models will seek to find temperature dependent conditions under which O_2 and H_2 environments can stabilize Pt NPs without forming metal oxide layers on them, as well as determining whether differently composed Pd and Rh NPs observe the same redox driven trade-off. In this study, energetic and structural data achieved using Density Functional Theory (DFT) modeling of M/γ-Al_2O_3 (M = Pt, Pd, Rh) systems will be applied to a multiscale Reactive Force Field (RFF) optimization approach, which will parameterize RFFs capable of performing Molecular Dynamics (MD) simulations. RFF MD simulations will then be completed under different temperatures, H_2/O_2 coverages, metal NP sizes, and other considerations to determine the redox reaction conditions under which the smallest possible metal NPs can retain catalytic activity while staying stabilized on γ-Al_2O_3 supports.

531846 A Theoretical Examination of Nitrogen Photofixation on Rutile TiO_2(110)

Abstract:

The photofixation of nitrogen is a promising reactive process for creating distributed production of nitrates and ammonia requiring only ambient conditions. The reaction has been demonstrated experimentally over titania catalysts as early as the 1970's, and can produce both oxidized and reduced compounds. Amazingly, this reaction proceeds at room temperature and pressure and only requires nitrogen, water, and light. This work has clear commercial potential, and beyond industry such benign process conditions would allow nitrogen products to be made in a more distributed way, decreasing the carbon emissions from transportation and facilitating improved management of fertilizer application. However, despite it's age and clear benefits, little fundamental work has been done on this system and currently only low reaction rates have been achieved. Gaining theoretical understanding of the mechanism of the reaction will enable optimization of the active site and mechanism of the process. In this talk we discuss the mechanistic challenges of nitrogen photofixation over titania, and present ab initio calculations for the screening of possible active sites based on a tradeoff of stability and reactivity. The potential roles of defect sites and impurities are also discussed, and we hypothesize that the process occurs over meta-stable active sites that are generated by reactions of impurities and photo-generated holes. These findings address several inconsistencies in the literature, and suggest new strategies for the rational design of more efficient nitrogen photofixation catalysts.

531836 Competitive Binding of Ethylene, Water, and Carbon Monoxide in Metal-Organic Framework Materials with Open Cu Sites

Abstract:

Metal-organic frameworks (MOFs) with open metal sites (OMS) are known to have selectivity in olefin/paraffin separations. One challenge associated with these separations is that other species that may bind strongly to OMS may also be present in feed streams, causing competition for these sites. We used Density Functional Theory (DFT) to assess the binding energy of ethylene, water and carbon monoxide on a set of more than- 60 MOFs with open Cu sites in the form of Cu dimers. One useful observation from our results is that the relative binding energies of pairs of molecules (e.g. ethylene and water) can be calculated accurately from calculations that hold the MOF structure rigid and only relax the positions of the adsorbing molecules. These kinds of calculations are far more numerically efficient than calculations that relax all degrees of freedom in the system, so this observation will be useful in future efforts to screen larger collections of materials. A second observation is that the binding energies of each molecule in the 60 MOFs are quite similar to the binding energies in CuBTC, an exemplar MOF with open Cu sites in the form of Cu dimers. Analysis of the variations that do exist in the binding energies among materials points to possible avenues for controlling either the absolute binding energies or the relative binding energies of species associated with OMS in these materials.

536375 Thermal Transport in Interpenetrated Metal-Organic Frameworks

Abstract:

Thermal Transport in Interpenetrated Metal-Organic Frameworks

Kutay B. Sezginel, Patrick A. Asinger, Hasan Babaei, and Christopher E. Wilmer

Department of Chemical and Petroleum Engineering, University of Pittsburgh, 3700 Oâ€™Hara Street, Pittsburgh, Pennsylvania 15261, United States

Metal-organic frameworks (MOFs) are crystalline materials comprising building blocks of inorganic nodes and organic linkers. The record high surface area and porosity of MOFs led researchers to investigate using MOFs as adsorbents for gas storage applications. In practice, the usefulness of metalâ€“organic frameworks (MOFs) for many gas storage applications depends on their ability to rapidly dissipate the heat generated during the exothermic adsorption process. MOFs can be precisely designed to have a wide variety of architectures which can allow tuning their thermal conductivity. While the focus for structureâ€“property relationships in MOFs has expectedly been gas-uptake capacity and selectivity, limited insight exists on the thermal transport characteristics of MOFs. Investigating design principles for thermal transport in MOFs is critical for allowing these materials to reach their full potential for gas storage and separations applications

In this work, we use molecular dynamics simulations to investigate the effect of interpenetration on the thermal conductivity of MOFs. We find that the addition of a parallel thermal transport pathway yields a thermal conductivity nearly the sum of the two constituent frameworks. This relationship holds for a variety of interpenetrating MOFs with different atomic masses and a wide range of interframework interaction parameters. We show that both the strength and range of interactions between constituent frameworks play a significant role in framework mobility as well as framework coupling which can result in deviation from this relationship. We propose a simple model to predict thermal conductivity of the interpenetrated framework based on thermal conductivities of individual frameworks and an interframework interaction parameter which can account for this deviation.

We believe these results provide important insights into the gas adsorption application of interpenetrated MOFs especially for small gases such as H_2. The inclusion of additional frameworks could increase storage capacity of the gas by increasing adsorption sites and provide additional thermal transport pathways which allows the generated heat to dissipate more quickly.

532818 QM and QM/MM Treatment of Ionic Liquid Binding to Cysteinated Porphyrin

Abstract:

Ionic liquids have the potential for driving industrial processes in a sustainable fashion due to their environmentally benign characteristics such as low vapor pressure, design flexibility and flammability. Although the negligible vapor pressure has potential to eliminate air emissions compared to conventional industrial solvents, the aqueous solubility of ionic liquids presents environmental concerns due to their toxicity. Thus biodegradation of ionic liquids is critical in the overall life-cycle assessment of these solvents. Experimental investigations have suggested that ionic liquids can be biodegraded under a variety of conditions; however, the molecular level explanations are currently lacking. To provide an insight into the biodegradation pathway, we select the ionic liquids 1-n-alkyl-3-methylimidazolium $[C_n mim]^+$ (n=2, 4, 6, 8, and 10) and cytochrome P-450 as the enzyme for their role in the oxidation of wide variety of molecules in aerobic and anaerobic environments. The catalytic site of the enzyme is modeled as an iron porphyrin molecule conjugated with the amino acid cysteine (FePCys). The ionic liquid-FePCys interaction is modeled using dispersion-corrected density functional theory to obtain binding energy. Components of the binding energy are obtained using the localized molecular orbital (LMO) energy decomposition analysis approach. As an enhanced description of the enzymesâ€™ active site, residues in the neighborhood of the central core were coupled in a QM/MM framework. Results from both the QM and QM/MM models will be compared to elucidate the effects of the ionic liquid substrates and relative orientations of ionic liquids and FePCys. Implications of these results in terms of ionic liquid biodegradation will be discussed.

527316 Computer-Aided Precursor Design and Process Development for Nanostructured Film Deposition

Abstract:

Applications of computational chemistry and molecular simulation to precursor design and process development for nanostructured film deposition will be discussed. Both Atomic Layer Deposition (ALD) and Chemical Vapor Deposition (CVD) are used to fabricate ultrathin and conformal thin film structures for many semiconductor and thin film device applications. ALD is particularly advantageous for integrated circuit manufacturing which requires control of film structure in the nanometer or sub-nanometer scale. One of the bottlenecks for detailed kinetic model building for product and process development is reliable thermochemistry and reactivity estimation. A detailed understanding of the microkinetics for the gas-phase and surface chemistries of precursor decomposition will allow for optimizing reactor design and process conditions. First-principles reaction path analysis of decomposition mechanisms can significantly decrease experimental costs through initial virtual screening of both potential precursors and surface cleaning requirements. Simultaneously, these studies can also improve the fundamental understanding of precursor stability and surface reactivity. In this collection of studies, the results demonstrate how theoretical investigations are currently impacting precursor discovery and deposition understanding in industry.

538685 Design of Biomaterials By Simulation and Experiment:Molecular Recognition, Assembly, and Applications

Abstract:

The development of functional biomaterials remains largely driven by serendipitous trial-and-error studies, whereby the rational understanding and design using modeling and simulation play an increased role thanks to more affordable computing resources and more accurate models. This talk describes simulation techniques and capabilities at the 1 to 1000 nanometer scale to understand recognition and assembly of metal, oxide, and biomineral nanostructures and their biomolecular interfaces. Applications to nanocrystal growth, catalyst design, hydrogels, and therapeutics are shown, including examples to use such methods as mainstream tools. Specific adsorption and assembly of peptides and macromolecules on metallic and oxide/hydroxide nanostructures will be described according to measurements and simulations, including predictions of chemical reactivity and guidance in designing new commercial products. Surface model databases and force fields in chemical accuracy such as the Interface force field appear to be critical as details of surface chemistry, electrolyte composition, and interfacial interactions play a key role.

533523 Fully Automated Molecular Design with Atomic Resolution for Desired Properties

Abstract:

Fully Automated Molecular Design with Atomic Resolution for Desired Properties

Hsuan-Hao Hsu, Chen-Hsuan Huang, Shiang-Tai Lin*

Department of Chemical Engineering

National Taiwan University, Taipei 10617, Taiwan

*Corresponding Author¡¦s E-mail: stlin@ntu.edu.tw

Keywords: genetic algorithm, COSMO-SAC, octanol-water partition coefficient, molecular design, simulated annealing.

Abstract

The value of fine and specialty chemicals is often determined by the specific requirements in their physical and chemical properties. Therefore, it is most desirable to design the structure of chemicals to meet some targeted material properties. In the past, the design of specialty chemicals has been based largely on experience and trial-and-error. However, recent advances in computational chemistry and machine learning can offer a new path to this problem. We applied the genetic algorithm (GA), based on the Darwin¡¦s theory of evolution and natural selection, combining with simulated annealing (SA) for this purpose. We have developed a new molecular data structure that allowed for efficient creation of new chemical species with GA. In this presentation, we demonstrate two successful examples where the structure of a chemical of specified value of octanol-water partition function and the energy gap (LUMO-HOMO) can be designed by computers. In prediction of K_{ow}, it consists of two parts, a robust method, the COSMO-SAC activity coefficient model, that predicts the activity coefficient with input of only the chemical structure, and the atom-based GA-SA. In the prediction of energy gap, we use AM1 method to calculate the energy gap for a chemical, and use GA-SA to search for the desired value of energy gap. We will show that the target value of K_{ow} and energy gap can be achieved within 1% of the target with a proper set of evolution parameters (including the size of population, the probability of selection, the rate of temperature annealing, etc.).

References

1. Sandler, S. I., *Chemical, Biochemical, and Engineering Thermodynamics*. 4th ed.; John Wiley & Sons, Inc.: New York, 2006.

2. Lin, S.-T.; Sandler, S. I., A priori phase equilibrium prediction from a segment contribution solvation model. *Ind. Eng. Chem. Res.* **2002,** 41, 899-913.

3. Zhang, J.; Peng, D.; Song, Z.; Zhou, T.; Cheng, H.; Chen, L.; Qi, Z., COSMO-descriptor based computer-aided ionic liquid design for separation processes. Part I: Modified group contribution methodology for predicting surface charge density profile of ionic liquids. *Chem. Eng. Sci.* **2017,** 162, 355-363.

4. Zhang, J.; Qin, L.; Peng, D.; Zhou, T.; Cheng, H.; Chen, L.; Qi, Z., COSMO-descriptor based computer-aided ionic liquid design for separation processes: Part II: Task-specific design for extraction processes. *Chem. Eng. Sci.* **2017,** 162, 364-374.

5. Peng, D.; Zhang, J.; Cheng, H.; Che, L.; Qi, Z., Computer-aided ionic liquid design for separation processes based on group contribution method and COSMO-SAC model. *Chem. Eng. Sci.* **2017,** 159, 58-68.

6. Lin, S.-T.; Hsieh, C.-M.; Lee, M.-T., Solvation and chemical engineering thermodynamics. *J. Chin. Inst. Chem. Eng.* **2007,** in press.

7. Venkatasubramanian, V.; Chan, K.; Caruthers, J. M., Computer-aided molecular design using genetic algorithms *Comput. Chem. Eng.* **1993**.

534077 Beyond United-Atom Lennard-Jones: Reliable Prediction of High Pressure Viscosities

Abstract:

Accurate prediction of viscosity (Î·) at high pressures (P) necessitates an extremely reliable force field for at least two reasons. First, the viscosity at a given density (Ï) is highly sensitive to the potential function form and associated parameters, especially those of the non-bonded interactions. Second, the viscosity depends strongly on the predicted density, which is also very sensitive to the force field.

As preliminary work suggests that traditional united-atom Lennard-Jones n-6 force fields are not capable of accurately predicting high pressure viscosities, a novel anisotropic-united-atom model is proposed. To develop a highly accurate force field, the CH_3, CH_2, CH, and C non-bonded parameters are optimized simultaneously using a large data set, consisting of PÏT and caloric properties over a wide range of state points, with particular emphasis at high pressures, for several normal and branched alkanes. Since the challenge compound is 2,2,4-trimethylhexane (TMH), we use 2,2,4-trimethylpentane (TMP, a.k.a. isooctane) as a surrogate molecule to refine the non-bonded parameters and, consequentially, to improve transferability.

An essential aspect of the challenge is to provide meaningful estimates of uncertainty. For this reason, uncertainties in the predicted TMH viscosity are quantified using three different methods. First, we estimate systematic bias in the force field by comparing the simulated and REFPROP Î· values for the surrogate compound, TMP, at the challenge temperature and pressures. Second, we account for the uncertainty in Î· that is associated with uncertainties in Ï for a given P. Third, we implement Bayesian inference to quantify and propagate the uncertainty in the force field non-bonded parameters.

538566 Prediction of the Viscosity-Pressure Relation Using Nonequilibrium Molecular Dynamics

Abstract:

Nonequilibrium molecular dynamics simulations provide an efficient way of calculating transport coefficients of molecular fluids. These methods have the advantage of being very efficient and accurate. They also provide a direct access to the response of the fluid when subjected to shear, leading to a characterization of the variations of viscosity as a function of the shear rate. Here we discuss the application of methods relying on the SLLOD equations of motion to the prediction of the viscosity-pressure relation of branched alkanes.

541570 Results of the Tenth Industrial Fluid Properties Simulation Challenge: Â Pressure-Dependence of Viscosity for a Short, Branched Alkane

Abstract:

The 10th Industrial Fluid Properties Simulation Challenge will test the capability of molecular dynamics simulation to provide the property of liquids most important to elastohydrodynamic lubrication (EHL), the pressure-viscosity relation. The temperature dependence at elevated pressure could be the subject of a future challenge.

537552 Open Chemistry and Jupyter: Platform for Data Mining and Machine Learning

Abstract:

Open, interactive interfaces employing data-centric workflows have been developed, reusing best-of-breed open source software in order to deliver an integrated platform for knowledge discovery. The Jupyter project offers a powerful electronic notebook interface, with server-side software kernels executing Python code. The JupyterLab web frontend provides a web-based interface with interactive cells where code can be edited, and panels where data can be visualized in various ways. Coupling these interfaces with a powerful data server, capable of triggering simulations, analyses, and other workflows offers a powerful capability to seamlessly execute codes from a pre-configured environment, store data, and apply machine learning techniques. The use of software containers improves both modularity and reproducibility.

Extension of the Python software kernels and web interface with chemistry specific capabilities results in a powerful software for chemical data. The open source platform will be described, along with links to a number of community codes from the computational chemistry community, machine learning, and the emerging computational chemistry machine learning community. The project is being developed in the open, with robust interfaces that facilitate the addition of new techniques. The core of the platform is a chemically aware data server, coupled with job submission capabilities, and data analytics.

The use of industry standard web programming interfaces, data formats, and modern HTML5 web components will be described. The use of the next-generation JupyterLab interface, its extension, and the integration of batch scheduling, search, and analytics will be covered. All major components will work in any modern web browser, including 3D visualization, analysis, and code execution. The use of Python kernels offers an ideal environment for analysis, reusing the tools already available in that ecosystem including significant investments in data mining and machine learning tools. A companion HTML5 application offers a simpler view of results that can be shared more widely, with full access to data, and linkage to the notebooks producing the data.

535198 Pythia: A Toolbox for Structural Analysis with Machine Learning

Abstract:

The recent explosion of interest and progress in machine learning (ML) methods has driven a proliferation of their application to soft matter systems. ML promises to deliver novel, automatic characterization techniques to solve previously insurmountable problems and it has already been successfully applied in several key areas for both disordered and ordered materials. However, researchers attempting to utilize ML methods on new systems often encounter challenges when the most appropriate data representation for their problem of interest is still unknown. To help alleviate this problem, we present Pythia, an open-source python library for generating numerical descriptions of particle configurations. Pythia allows users to select from a palette of descriptors ranging in complexity from simple to sophisticated. We demonstrate how Pythia can be combined with standard ML methods to identify complex structures, create phase diagrams, analyze crystal grains, and moreâ€"all in a high-throughput manner.

532141 Moleculenet: A Benchmark for Molecular Machine Learning

Abstract:

Molecular machine learning has been maturing rapidly over the last few years. Improved methods and the presence of larger datasets have enabled machine learning algorithms to make increasingly accurate predictions about molecular properties. However, algorithmic progress has been limited due to the lack of a standard benchmark to compare the efficacy of proposed methods; most new algorithms are benchmarked on different datasets making it challenging to gauge the quality of proposed methods. This work introduces MoleculeNet, a large scale benchmark for molecular machine learning. MoleculeNet curates multiple public datasets, establishes metrics for evaluation, and offers high quality open-source implementations of multiple previously proposed molecular featurization and learning algorithms (released as part of the DeepChem open source library). MoleculeNet benchmarks demonstrate that learnable representations are powerful tools for molecular machine learning and broadly offer the best performance. However, this result comes with caveats. Learnable representations still struggle to deal with complex tasks under data scarcity and highly imbalanced classification. For quantum mechanical and biophysical datasets, the use of physics-aware featurizations can be more important than choice of particular learning algorithm.

535871 Towards an Open-Source Implementation of Spatially-Resolved Molecular Fingerprinting Methods in Machine Learning-Based Predictions of Material Properties

Abstract:

The fields of artificial intelligence (AI) and machine learning (ML) have experienced tremendous growth and success over the past few decades. This has led to the development of innovative and powerful solutions to a host of problems in various fields including robotics, medicine, manufacturing, etc. AI and ML also show great promise for applications in materials design and characterization. One such possibility is to help uncover characteristic structure-property relationships in various classes of materials. An important roadblock towards ML-based direct mapping between structures and properties is the availability of numerical representations that can capture various structural signatures while providing spatial resolution and being invariant to operations such as geometric rotations and translation. Cartesian coordinates traditionally used to represent molecular structures suffer from a lack of uniqueness and variance under such property-preserving geometric operations. Over the past decade or so, progress has been made in developing alternative representations, called "fingerprints," that are suitable for ML applications. In this work, we review and compare fingerprints that provide spatial resolution and propose a new classification scheme to better grasp their applicability and limitations. Moreover, we introduce a new open-source software package called SEING, designed and implemented for easy computation of spatially resolved molecular fingerprints. We finish by showing example applications for force prediction in crystals such as Si, Al and W with a generalization accuracy on the order of 0.1 meV $Å^{-1}$.

536223 Bayesian Optimization of Molecular Structures: Data-Driven Sampling for Molecular Conformers

Abstract:

While most molecules with more than 4 atoms have multiple conformers as local minima, generating accurate ensembles requires of both efficient sampling and accurate potential energy surfaces. Despite many existing methods to find accurate molecular conformations, each year novel methods are published, indicating the level of difficulty in the problem. We will briefly review the history of the field, including previous data-driven approaches, as well as modern repositories, including both experimental and computational data. We will discuss recent efforts to benchmark force fields, semiempirical quantum chemical methods, density functional approximations, and higher-level quantum methods for conformer evaluation. Rankings and correlations between low-level methods are amazingly poor. Discussion will focus on new data-driven methods to efficiently sample the multi-dimensional problem and accurately provide energy evaluations.

532043 End-to-End Learning for Prediction of Optoelectronic Properties of Organic Photovoltaic Polymers

Abstract:

Despite algorithmic improvements to quantum-mechanical methods, computational design of atomistic systems is limited by the vast combinatorial space of possible molecular structures. Algorithms that efficiently interpolate more detailed simulations across this combinatorial landscape are therefore needed. Traditional approaches to building quantitative structure-property relationship models cast molecules to a fixed-size numeric vector using static methods that do not change during the learning process. In this talk, we demonstrate how end-to-end learning approaches [1] drastically improve prediction accuracy over fingerprint-based approaches for large datasets. We apply the method to a computational database for active-layer materials for organic photovoltaic solar cells. The database contains approximately 30,000 unique chemical compounds, for which optoelectronic properties were calculated using time-dependent DFT with the B3LYP/6-31g(d) model chemistry. Our method is able to predict HOMO, LUMO, and band gap values extrapolated to the polymer limit [2] using only the monomerâ€™s structural information. Prediction accuracy is shown to have a root mean squared error of approximately 0.05 eV for HOMO, LUMO, and band-gap calculations while taking many orders of magnitude less time to perform. Additionally, we demonstrate the ability of the algorithm to learn on subsets of the data, and the amount of data required to fully parameterize end-to-end learning algorithms such that they outperform traditional fingerprinting-based methods.

[1] Gilmer J, Schoenholz SS, Riley PF, Vinyals O, Dahl GE. Neural Message Passing for Quantum Chemistry. arXiv. 2017;cs.LG. (2017).

[2] Larsen RE, J. Phys. Chem. C, 120, 9650-9660 (2016).

536780 Practical Applications of Machine Learning to Catalyst Design and Discovery

Abstract:

Heterogeneous catalysis is fundamental to chemical industry and consumes several percent of the entire global energy supply. Reducing this usage and enabling next-generation energy solutions such as direct conversion of CO2 to fuels requires the design of new catalysts with optimal activity, selectivity, and stability. Scientific computing advances have enabled electronic structure codes to aid in this design process but fundamental limitations make it unlikely that direct simulation of macroscopic catalysts will be possible. The huge design space can be reduced by recognizing similarities in materials (developing structural fingerprints) and adopting regression tools from the systems engineering or machine learning communities to provide useful surrogate models as a guide for full-accuracy calculations. I will discuss how we approach the coupled problems of database generation, calculation automation, and active learning to search for intermetallic catalysts with near-optimal properties. We distinguish these approaches from typical high throughput screens because the final calculations are functionally identical to the standard computational chemistry workflow that would be used for individual surfaces. We demonstrate the successful identification of more than 100 new candidates for the CO2 reduction reaction, and show that the final predictive models can aid in the determination of the active facet in polycrystalline catalysts.

523452 Mapping Configurationally-Dependent Electronic Structure to Coarse-Grained Models with Machine Learning

Abstract:

The modeling of functional soft materials involves the synthesis of multiscale simulation techniques to describe the impact of nuclear degrees of freedom on the electronic structure of the material. For these systems, configurational sampling is often a substantial computational bottleneck warranting the use of computationally-cheaper coarse-grained models. However, these coarse simulations must be followed by atomistic backmapping, as well as electronic structure calculations on each configuration, in order to accurately describe the material's electronic structure. Here, we present a simulation protocol for mapping configuration-dependent electronic structure directly to coarse-grained nuclear degrees of freedom using techniques from supervised machine learning. We describe the impact of this approach on the acceleration of simulations, as well as the potential ability to discover coarse-grained representations relevant to organic semiconductors.

537884 Role of Pore Chemistry and Topology in the CO_2 Capture Capabilities of MOFs: Molecular Simulation and Machine Learning

Abstract:

Open-framework materials (OFMs) such as metal-organic frameworks (MOFs) are promising for applications such as CO_2 capture. The challenge is that with millions of possible "crystal topology + composition" combinations it is very difficult to identify the best possible MOFs, to clearly elucidate MOF performance limits, and to isolate the impact of different MOF features on performance solely doing experiments. High throughput computational screening (HTCS) have emerged as a powerful tool to study large numbers of MOFs. However, despite vast amounts of CO_2 adsorption data collected in various HTCS efforts, the role of the MOF pore chemistry and topology remains to be clearly elucidated due to factors in earlier HTCS studies such as i) limited number of topologies considered, ii) limited number of functional groups considered, iii) incomplete MOF series, and iv) accuracy uncertainties for the description of electrostatic interactions.

In this work, we leverage our tools to construct MOFs using the TobaCCo code[1] and to inexpensively calculate charges that describe well electrostatic interactions using MBBB charges[2] to study the adsorption of CO_2 and CO_2/N_2 and CO_2/H_2 mixtures in a "complete" population of MOFs. The "complete" population of MOFs contains all possible "crystal topology + functionalization" combinations for 15 topologies and 13 functionalized building blocks, which has allowed us to comprehensively examine the role of pore topology and chemistry on the CO_2 capture capabilities of these open framework materials. Then, we use this knowledge to provide "intuitively derived" descriptors based on data from DFT calculations and GCMC simulations to train machine learning algorithms to quantitatively predict CO_2 capture properties of the studied materials without the need of adsorption simulations.

We evaluated the performance of six different machine learning algorithms, including artificial neural networks (ANNs) and gradient boosting machines (GBMs). We found that, for similar computational cost, gradient boosting machines consistently outperformed neural networks in prediction accuracy. Finally, by using individual conditional expectation (ICE) plots we were able to further analyze the response of CO_2 capture "metrics" to changes in MOF features, including chemistry and topology, allowing us to predict the optimal values of these features for future optimization of MOF design.

References

1. Cryst. Growth. Des. 2017, 17 (11), pp 5801-5810
2. J. Chem. Theory Comput., 2018, 14 (1), pp 365–376

535428 Accelerating the Characterization and Design of Nanoporous Materials with Data-Driven Models

Abstract:

Metal-organic frameworks (MOFs) are a class of nanoporous materials with highly tunable pore shape and chemistry. They are synthesized via self-assembly of metal nodes connected by organic linkers to form crystalline porous materials which have been explored for diverse applications including gas storage and catalysis. High-throughput molecular simulations can identify promising materials out of thousands of candidate MOFs, but the interrogation of this vast amount of generated data has been only rudimentary in the literature. We want to use more sophisticated techniques to glean more insights from this data. In this work, we applied machine learning methods to accelerate high-throughput computational screening of MOFs and to develop better structure-property relationships. Our approach uses least absolute shrinkage and selection operator (LASSO) regression to extract insights from MOF energy landscapes, which are rapidly calculated using a probe molecule. The model is highly accurate, interpretable, robust, and three orders of magnitude faster than detailed molecular simulations. To demonstrate the usefulness of this model, we screened 55,000 MOFs and identified a candidate MOF with high hydrogen storage capacity, which we also confirmed experimentally. The approach can be easily applied to other classes of porous materials and adsorbate molecules such as methane.

535096 Deformation and Yield in Semicrystalline Polymers

Abstract:

Many polymers are heterogeneous in the solid state, comprising both ordered and disordered domains with nanometric dimensions that are intimately related through long chain molecules that traverse from one domain to another. Semicrystalline polymers as well as many copolymers such as thermoplastic polyurethanes fall into this category. Modeling the thermomechanical properties of these materials is complicated by the need to capture not only the behavior of the ordered and disordered domains, respectively, but also their coupled response through the interface. To study such materials with atomistic resolution, we use the Lamellar Stack Model [1], which serves as the elementary building block for more complicated morphologies. By studying the more tractable problem of molecular deformation mechanisms within the lamellar stack using molecular dynamics simulations, one can begin to understand the processes operative throughout the complicated, heterogeneous morphology.

Results are reported for atomistic models of semicrystalline linear polyethylene (PE) and phase-segregated thermoplastic polyurethanes (TPU) containing amorphous polytetramethylene oxide (PTMO) soft segments and 4,4'-diphenylmethane diisocyanate (MDI) hard segments with n-butanediol chain extenders. The Lamellar Stack Model was constructed in each case using the Interphase Monte Carlo (IMC) procedure [2,3] to equilibrate both the packing and topology. Following equilibration, representative ensembles of configurations were then transferred to LAMMPS [4] for re-equilibration in the NPT ensemble, followed by deformation at constant strain rate to large strain, typically under uniaxial strain (constant lateral dimensions) or uniaxial stress (constant lateral stress) conditions.

The response of the lamellar stack was found to depend on both the mode and rate of deformation. In semicrystalline PE, mechanisms of response were categorized as fine crystallographic slip, occurring at low strain, followed by either cavitation in the amorphous domain or melting/recrystallization of the crystal domain [1,5]. The signature of melting/recrystallization behavior in particular was a rise in stress up to the yield point, followed by a sudden drop in stress as material was transferred from one domain to the other; under appropriate simulation conditions, this mechanism would occur repeatedly, giving rise to a cyclic rise and fall in stress with increasing strain. These events were found to be uncorrelated between configurations, resulting in the appearance of plastic flow post-yield after ensemble-averaging [6]. In the TPU case [7], under uniaxial strain deformation, cavitation of the soft domain was universally observed, in close analogy to semicrystalline PE. Chain pull-out associated with bridges and bridging entanglements was also observed. In contrast to semicrystalline PE, shear band formation and localized block slip mechanisms were found. In this case, the different mechanisms could be traced to the soft domain topology, but also the sliding of hydrogen-bonded sheets within the hard domain. The occurrence of these several mechanisms depends uniquely on the mechanical coupling between hard and soft domains in heterogeneous polymer solids, as represented by the disordered domain topology.

[1] Kim, J.; Locker, C.R.; Rutledge, G.C. *Macromolecules* **2014,**47 (7), 2515-2528.

[2] Balijepalli, S.; Rutledge, G.C. *J. Chem. Phys.* **1998,**109 (16), 6523-6526.

[3] in 't Veld, P.J.; Hutter, M.; Rutledge, G.C. *Macromolecules***2006,**39 (1), 439-447 DOI: 10.1021/ma0518961.

[4] Plimpton, S. J.*Comput. Phys.* **1995**, 117, 1–19.

[5] Lee, S.; Rutledge, G.C. *Macromolecules* **2011,**44 (8), 3096-3108.

[6] Yeh, I.-C.; Andzelm, J. W.; Rutledge, G. C. *Macromolecules***2015**, *48*(12), 4228–4239.

[7] Zhu, S., Lempesis, N., in 't Veld, P.J., Rutledge, G.C., Macromolecules 2018, 51(5), 1850-1864.

528177 Computing Mechanical Properties of Elastomers Under Multiaxial Deformation Using Molecular Modeling

Abstract:

In order to develop a better understanding of the impact of microscopic changes on macroscopic properties of elastomers, this work focuses on developing novel numerical methods for evaluating changes in elastomer chain conformations under multiaxial deformation. Random elastomer chain conformations were generated under unperturbed conditions using Floryâ€™s Rotational Isomeric State (RIS) Approach. After the unperturbed state of chain ensembles was analyzed through size and shape studies, the deformed state of the ensembles was explored through molecular modeling techniques to quantify external stresses. Uniaxial deformation, equibiaxial deformation and shear were applied to unperturbed chain ensembles by computing changes in the probability density distribution of chain end-to-end vectors and elastic free energy. The approach for computing changes in elastic free energy involved developing a probability distribution-based numerical method that can be applied across multiple modes of deformation. In order to determine the accuracy of the numerical method, it was initially applied to generated Gaussian chains and compared against known analytical equations. Then the numerical model was extended to computing elastic free energy change, force, and stress on RIS cis- and trans-1,4-polybutadiene chains. Compression forces were much greater than tension forces. Equibiaxial and uniaxial stresses in the same direction of extension were equal in magnitude and greater in magnitude than shear stress. Forces and stresses increased with deformation and showed expected dependences on chain volume and temperature. Significant variation was observed in moduli with number of chain repeat units, and the expected linearly proportional variation was observed with temperature. Chains of fewer repeat units correspond to smaller molecular weights between cross-links, leading to a more tightly cross-linked network with larger moduli than that of chains of more repeat units. Young's and shear moduli computed from the numerical model were in good agreement with available experimental results. Implementation of the numerical model along with the RIS method allows for the ability to include effects of various surface chemistries and end groups on conformation distribution to facilitate the study of filled systems such as rubber elastomers used in tires.

521190 Molecular Modeling of Mechanical Properties of Semicrystalline Polymer Fibers

Abstract:

Polymer fibers have various chemical and physical structures that affect their mechanical properties. Polymer chemistry, molecular weight, crystallinity, and orientation in a polymer fiber are factors that influence mechanical strength and deformation behavior. We carried out molecular dynamics (MD) simulations using a united atom force field to model semicrystalline polymers, including polyethylene. From the MD simulations, we estimated the mechanical properties, and compared them to experimental data for validation. Data such as stress-strain curves can be used to fit parameters for constitutive equations. Developing molecular modeling strategies for accurately capturing mechanical properties allows for parameterizing macroscopic mechanical models based on polymer fiber structural properties.

531996 Predicting Nematic Coupling of Polybutadiene Using Atomistic Molecular Dynamic Simulations

Abstract:

Various experiments suggest the presence of nematic interactions in flexible polymer melts and networks. These interactions aï¬€ect how these polymers respond to external fields like stress fields, magnetic fields and electric fields, even when liquid crystalline phases are not apparent. Such interactions can be quantified by nematic coupling constant, and these interactions affect the molecular ordering in polymer systems. To determine the value of the nematic coupling constant in flexible polymers, we combine self consistent field theory (SCFT) with atomistic molecular dynamic simulations of chains under external tension in an isotropic melt; we have previously used this approach to predict the nematic coupling parameter for a semiflexible polymer, poly(3-hexylthiophene-2,5-diyl). We apply this method to determine the nematic coupling constant of poly-butadiene (PB) and determine the scalar order parameter for a melt of strained PB chains and for free PB chains embedded in such a strained melt. The ratio of these order parameters at various temperatures is then compared with the experimental results. The ratio obtained using our method and extrapolated to experimental conditions predicts a value of 0.8 and is consistent with experimental results (~0.9). Our simulations corroborate theory by indicating that coupling between orientation of neighboring segments exists in PB.

534629 A Multiscale Modeling Approach to Characterizing Structural and Transport Properties in Diblock Copolymer Polymerized Ionic Liquids

Abstract:

We utilized a multiscale modeling approach to explore ion structure and transport in diblock copolymers of 50wt% poly(1-butyl-3-vinylimidazolium)-hexafluorophasophate and polystyrene. We employed dissipative particle dynamics (DPD) with charge smearing on a coarse-grained system. Like and unlike group interaction parameters were thermodynamically derived. Intramolecular potentials were manually varied until bond and angle distributions matched those observed from atomistic molecular dynamics (MD) simulations. A phase-segregated atomistic configuration was obtained using an inverse coarse-graining procedure on the final DPD configuration. We analyze trends in diffusivity and ion-association structure and dynamics using atomistic MD. We focus on the locations of ions, such as near boundaries and within bulk phases, as a key influencer of polymer dynamics and ion mobilities.

538855 Applying Protracted Colored Noise Dynamics to Dramatically Increase the Simulation Efficiency of Linear Polymer Systems

Abstract:

Molecular dynamics simulations of polymers are significantly limited because of the large time and length scales required by the viscous nature of polymeric systems. A previous stochastic simulation method called Protracted Colored Noise Dynamics (PCND) increased the local diffusivity of highly viscous molecular systems by applying correlated stochastic forces (colored noise). This method dramatically increased the diffusivity of a molecular glass and resulted in a crystallization rate that was approximately three orders of magnitude faster than normal molecular dynamics systems. We have applied PCND to linear polymer systems by directing the stochastic colored noise along the backbone of the polymer contour. This increases the reptation mode of the polymer, and hence the local diffusivity. We tested the effectiveness of this new method using simulations of defect annealing in lamellae formed by the directed self-assembly of block copolymers. This polymer PCND increased defect annihilation by approximately four orders of magnitude compared to molecular dynamics. Application of molecular dynamics restores the correct domain size, indicating that the free energy landscape is not being significantly altered. The computational overhead associated with polymer PCND is minimal, and outweighed by the increased equilibration rate. We also studied the effective parameter space of the time relaxation of the colored noise and the magnitude of the colored noise. If these parameters are too large, the stochastic forces overwhelm the system and do not produce equilibrated structures. The effectiveness of the method is due, in part, to a strategic violation of the fluctuation dissipation theorem. This does perturb the dynamic trajectory of the simulation, but correct structures are produced by switching off the stochastic force at the end of the simulation. This makes this application of PCND very effective at accessing polymer conformation space that contains significant energy barriers.

528709 An Atomistic Evaluation of the Compatibility and Plasticization Efficacy of Phthalates in Poly(vinyl chloride)

Abstract:

Plasticizers are widely used to improve the properties and processability of certain polymer materials, such as polyvinyl chloride (PVC), in the plastic industry. Phthalates are the most widely used plasticizers on the market. These compounds tend to migrate out of the host polymer owing to their lack of chemical bonding and poor thermodynamic affinity with the polymer matrix, which results in the deterioration of materials performance, environmental contamination, and even harm to human health. Increasing governmental regulation is driving the search for more green and stable alternatives. Industrial selection of plasticizers largely relies on semi-empirical rules and even for successful candidates, it is often difficult to answer why they outperform other molecules in terms of effectively reducing the glass transition temperature and elastic modulus of the material. In this study, we build full-atom molecular models for the reliable prediction of plasticizer performance based on its chemical structure. A group of phthalates with similar molecular structure but different side chain configurations are simulated as test cases. Simulation results are in good agreement with experimental reports of their plasticization efficiency, thermodynamic compatibility, and migration resistance. Molecular insight from our simulation also helps us understand the fundamental relationship between plasticizer structure and its performance. This is the first step towards our eventual goal of developing a computational protocol for screening candidate plasticizers based on their chemical structures and finding high-performance and stable alternatives to phthalates. Beyond the PVC-plasticizer systems, we also note that chain length has a nontrivial effect on the cohesive energy and solubility parameter of long-chain polymers. This dependence will be discussed and should be an important consideration in the thermodynamic property prediction of polymers from molecular simulation.

538586 Zwitterionic Contribution to the Hydration Lubrication Dynamics of Poly(2-methacryloyloxyethyl phosphorylcholine)

Abstract:

Poly(2-methacryloyloxyethyl phosphorylcholine), or pMPC, is a biocompatible polymer shown to exhibit remarkable tribological properties and has received extensive experimental study as a possible artificial synovial fluid [1], especially as a lubricant for orthopedic implants [2]. These studies have attributed the lubricity of pMPC to a hydration lubrication [3] mechanism, but insight related to the underpinnings of this mechanism at the atomic scale is still lacking. In previous work [4,5], we showed that the polymer's zwitterionic character, specifically the charged phosphoryl and choline groups, are integral to the hydration layers that form. This is due to a combination of the polymer's natural affinity for hydration (i.e. hydrophilicity), and more subtly, the inter-monomer phosphoryl-choline interactions which aid in "unfurling" the polymer structure, allowing for increased hydration of the phosphoryl groups. Although our previous efforts have been instrumental in determining the atomic contributions to the hydration lubrication mechanism, our understanding of the intermonomeric effects that assist this mechanism (monomer-water hydrogen bonding, monomer-monomer hydrogen bonding, etc) is still lacking.

In this work we explore the hydration structure and dynamics of pMPC polymers of varying chain lengths, pMPCn, in solution using molecular dynamics simulations. System setup and parameterization was aided by the open source Molecular Simulation and Design Framework6 (MoSDeF), developed at Vanderbilt. We compare bulk solvated pMPCn to poly(2-methacryloyloxyethyl phosphate)n (pMPn) to quantify the effect that the choline group has on hydrogen bonding between the pMPC polymer and water. When the choline group is removed, total hydrogen bonding between the pMP polymer and water is found to be reduced by approximately a factor of 2. The choline group plays an essential, but unintuitive role in the stabilization and increased hydration of the polymer through inter-monomer hydrogen bonding of the choline-phosphoryl groups. Although these interactions between the phosphoryl and choline groups limit possible water-phosphoryl hydrogen bonding, the "unfurling" effect these interactions have on the polymer exposes more hydration sites overall, leading to the pronounced hydration lubrication mechanism observed.

References

[1] Chen, M., Briscoe, W. H., Armes, S. P., & Klein, J. (2009). Lubrication at Physiological Pressures by Polyzwitterionic Brushes. *Science*, *323*(5922), 1698–1701. https://doi.org/10.1126/science.1169399

[2] Kyomoto, M., Moro, T., Saiga, K., Hashimoto, M., Ito, H., Kawaguchi, H., ... Ishihara, K. (2012). Biomimetic hydration lubrication with various polyelectrolyte layers on cross-linked polyethylene orthopedic bearing materials. *Biomaterials*, *33*(18), 4451–4459. https://doi.org/10.1016/j.biomaterials.2012.03.028

[3] Klein, J. (2013). Hydration lubrication. *Friction*, *1*(1), 1–23. https://doi.org/10.1007/s40544-013-0001-7

[4] Klein, C., Iacovella, C. R., McCabe, C., & Cummings, P. T. (2015). Tunable transition from hydration to monomer-supported lubrication in zwitterionic monolayers revealed by molecular dynamics simulation. *Soft Matter*, *11*(17), 3340–3346. https://doi.org/10.1039/C4SM02883J

[5] W.L. Roussell, C. Klein, C.R. Iacovella, P.T. Cummings, C. McCabe, Molecular Origins of the Ultra-Low Friction Exhibited by Biocompatible Zwitterionic Polymer Brushes *Young Scientist*, May 2015

[6] "MoSDeF" [Online]. Available: https://github.com/mosdef-hub.

530430 Protein Stabilization in Non-Native Solvents with Random Copolymers

Abstract:

The ability to maintain the enzymatic activity of proteins out of their native aqueous medium is crucial for pharmaceutical and chemical processing industries. In foreign media such as organic and highly polar solvents, however, proteins are prone to a substantial reduction in their functionalities as their structure, solubility and conformational mobility are heavily affected. Recently, Panganiban et al. [1] demonstrated that a special case of random heteropolymers with four different monomers allow for highly effective stabilization of numerous enzymes in toluene. In this talk, we demonstrate via multiscale modeling and simulations how random copolymers with solvophobic and solvophilic groups encapsulate numerous proteins in different solvent conditions. We determine the key factors that govern the protein surface coverage of the random copolymers such as the adsorption energy, average copolymer composition and solvent selectivity. We find that the adsorbed polymer chains have considerably similar sequences, indicating that the proteins are able to select
certain copolymer sequences that best reduce their exposure to the unfavorable solvent. The results herein set the stage for computational design of random copolymers for stabilizing and delivering proteins across multiple media.

References

1. Panganiban et al. Science 2018, 359:1239-124.

*The research was supported by the Department of Energy Award No. DE-FG02-08ER46539.

527958 Knowledge Extraction Via Machine Learning from High-Throughput Catalytic Experiments

Abstract:

High-throughput experimentation on catalytic reactions produce a wealth of information. This quickly becomes problematic for researchers, who are tasked with examining this wealth of information and extracting fundamental chemical information to use for catalyst optimization. Recent advances in artificial intelligence, specifically in the field of machine learning, have produced automated methods of data analysis to quickly extract important relationships from complex datasets, allowing the researcher to focus on applying expert knowledge rather than crunching numbers.

This machine learning methodology has been applied to analyze a high-throughput screening of ammonia decomposition catalysts. Ammonia decomposition has been extensively studied in a fairly narrow design space predominately focused on Ru catalysts supported by single alkali metals[1]. In an effort to expand the design space, high-throughput experimentation was employed to study a wide variety of catalysts with differing elements (Ru-M-K, where M is a supporting metal) and weight loadings. They were tested for activity towards ammonia decomposition under varying reaction conditions, such as inlet ammonia concentration and reactor temperature. Several hundred potential catalyst features were chosen and a machine learning algorithm was utilized to extract catalyst features that caused the greatest change in catalytic activity. This knowledge was then used to guide further catalyst synthesis.

[1] S. Mukherjee, S. V. Devaguptapu, A. Sviripa, C. R. F. Lund, and G. Wu, â€œLow-temperature ammonia decomposition catalysts for hydrogen generation,â€ *Appl. Catal. B Environ.*, vol. 226, no. December 2017, pp. 162â€“181, 2018.

530541 Catalyst Characterization from Complex Infrared Spectroscopy: A Machine Learning Approach

Abstract:

Adsorbate vibrational excitations are a unique descriptor of adsorbate/surface interactions with the benefit of being relatively independent of temperature and pressure effects. The infrared spectra (IR) associated with activating adsorbate vibrational modes is very accurate, captures details of most vibrational modes, and can be obtained operando. Despite the possibility of obtaining detailed catalyst structural information and adsorption site preference throughout the course of a reaction from IR spectroscopy, the ability to interpret the resulting complex spectra is lacking. Current techniques depend on heuristic peak assignments and can only be done for relatively simple spectra. Quantitative assignment of adsorption sites can only be done on simple surfaces where the structure is known a-priori and requires concurrent measurements from mass spectrometry (MS) and low energy electron diffraction (LEED) studies at ultra-high-vacuum (UHV). We present a machine learning approach to determine adsorption site preference and local surface structure where complexity in the IR spectra is preferred. We combine first-principles calculations of carbon monoxide on platinum nanoparticles and a forward surrogate model to simulate IR spectra for training an inverse surrogate machine learning model. We test the machine learning model on both synthesized spectra generated from our forward surrogate model and on experimental spectra. This work demonstrates the ability to gain relevant information from experimental data using theory and computation.

538082 Heterogeneous Catalysis Kinetic Characterization Via Sparse Graphs

Abstract:

Kinetic rate estimation is essential to understanding the underlying elementary behavior of a catalyst. While computational kinetic methods are necessary for estimating an unknown system, there may be drawbacks in sensitivity to the selection of the reaction model, estimates of parameter conditions and inclusion of experimental results. Combining the use of physical models and experimental transient data, accurate kinetic rate information can be obtained with no assumption on the reaction mechanism while reducing the assumptions required on the experiment. We leverage data science and statistical techniques in the form of Gaussian Graphical Models to extend existing methodology in predicting kinetic rates. This methodology is applied to highly dense transient data from the Temporal Analysis of Products (TAP) reactor allowing for determination of the kinetics in an intrinsic manner within time as well as between experiments. The analysis results in a series of connected graphs to make better insights of the catalyst behavior as the surface evolves.

526981 Bayesian Experimental Design and Mean Field Microkinetic Modeling of Heterogeneous Catalytic Systems

Abstract:

Reaction kinetics experiments are valuable in providing mechanistic insights of a catalytic system, however, they can be time-consuming and expensive and may not be able to discriminate between plausible reaction pathways. Microkinetic models, using ab initio energetics, can in principle provide information of a catalytic system a priori; however, the intrinsic errors in modern ab initio methods (such as density functional theory) can result in significant errors in prediction of rates (a factor of 10 â€" 100 and sometimes higher). We propose a modeling framework that integrates experimental kinetic data and microkinetic modeling, to sequentially identify and select the most â €œinformativeâ€ data to build a robust mechanistic model.

Our framework uses a Gaussian Process (GP) model to quantify the uncertainty in energies of the chosen functional (e.g. PW91) by benchmarking with experimental adsorption energies, and propagates the errors through a microkinetic model formulated using DFT energies using the same functional. Then, Bayesian experimental design is employed to rank order candidate experimental conditions, and sequentially pick informative experiments to refine the model through Bayesian inference.

In this talk, the formulation of the GP model, the error propagation, and the experimental design strategies will be presented and discussed in the context of an illustrative example involving the low temperature water gas shift (WGS) reaction on copper.

531580 Thermochemistry of Gas-Phase and Surface Species *Via* Lasso-Assisted Subgraph Selection

Abstract:

First-principles modelling of chemical kinetics provides key insights into reaction mechanisms and better catalysts. However, such modelling becomes prohibitive for large reaction networks due to the computational burden of *ab initio* methods. This computational burden can be circumvented using graph theory-based semi-empirical methods, such as group additivity. These models assume that the thermodynamic properties of molecules linearly correlate with the frequency of graphs appearing in each molecule. However, the selection of these graphs and correction terms are heuristic, resulting in a suboptimal model.

To improve the accuracy of these methods, we explore the Least Absolute Shrinkage and Selection Operator (LASSO), an automatic descriptor selection algorithm. Graphs in each molecule are exhaustively enumerated, and LASSO regression is performed to find the most optimal set describing the enthalpy of formation. We have applied this framework to a variety of data sets. First, we gather heat of formation data from the NIST and the BURCAT datasets, and found that the LASSO-assisted graph selection shows superior performance over the heuristic-based group additivity (5.7 vs. 9.2 kcal mol^{-1} mean absolute error). Second, we apply this framework to the QM9 dataset and find our model performs comparable to the state-of-the-art machine learning models with a mean absolute error of 1.39 kcal mol^{-1}. Third, we apply this framework to lignin monomer adsorbates on Pt(111) and obtain a model with a mean absolute error of 2.08 kcal mol^{-1}, showing promising performance for surface species. We discuss how LASSO is especially useful for surface adsorbates to overcome the difficulty in manual identification of strain corrections.

528599 Theoretical Investigation of the Pt Catalyzed Hydrodeoxygenation of Succinic Acid to 1,4-Butanediol

Abstract:

The surface reaction network of the hydrodeoxygenation (HDO) of succinic acid (SUCC) to 1,4-butanediol (BDO) exhibts enormous complexity with hundreds of intermediates and thousands of possible elementary reaction steps. To investigate all theoretically possible elementary reactions at DFT/BEEF-vdW level of theory is not only cubersome, but likely also a waste of computational resources, considering that most likely only a small part of the total reaction network is of kinetic relevance. We present here a framework for optimizing the reaction network using models that are trained on the fly.

Our model framework consists of a database of reaction intermediates and transition states from both this and our previous studies on the HDO of propanoic acid, stacked Gaussian processes (GP) that predict both the unmeasured adsorption energies and transition state energies, and a mean-field microkinetic model which can identify key reaction intermediates, rate controlling steps and dominant catalytic cycles. The model is iteratively used until all relevant reaction intermediates and transition states are explicitly calculated at DFT/BEEF-vdW level of theory.

We applied this modeling framework to the HDO of SUCC to BDO on Pt (111) and Pt (100) and identified the dominant reaction pathway. Next, we predicted the uncertainties of quantities of interest such as the TOF to identify the active site. Finally, a comparison between our predicted values with experimental measurements is made to prove the applicability of our model.

534940 Large-Scale Exploration of Perovskites for Oxygen Evolution Via Adaptive Machine Learning

Abstract:

Perovskites with a ABO3 structure have shown great promise for catalyzing the thermodynamically and kinetically unfavorable oxygen evolution reaction (OER) in the (photo)-electrochemical systems. While it is very time consuming and costly to search for highly optimized catalysts by either empirical testing or the quantum-chemical calculations. In recent years, we have seen a dramatic rise of the use of machine-learning techniques for the materials design[1–3]; however, the biggest challenge of a machine-learning approach is the development of robust feature representations for the materials of interest and the accessibility to large training data sets.

In this regard, we introduce a novel electronic-structure descriptor, namely DOS entropy index, for representing the perovskite in the machine-learning models. The DOS entropy index characterizes the dissimilarities of the perovskites' complex electronic structures according to the pairwise Kullback-Leibler (KL) divergence of the associating atomic orbital-wise projected density of states (PDOS). Based on the selection of reference systems, the DOS entropy index has two generalized forms, i.e., atomic pairwise entropy index and entropy eigen-spectrum. Starting from a limited ~20 perovskites' DFT-calculated properties and the corresponding OER activities from the literatures, the Bayesian neural network optimized by the DOS entropy index is able to inference above 1000 double perovskites' OER activity likelihoods and the associating prediction uncertainties. In practice, our belief towards the model prediction accuracy can be recursively improved by validating the predicted materials with high uncertainties in the actual experiments and incorporate the validation information into the training sets. This adaptive learning framework shows great value in the practical applications for the large-scale exploration of perovskites' measured OER activities using the easily accessible computational descriptors. Furthermore, the quantified uncertainties by the Bayesian rules also allow a potential improvement of the model's generalizability with a minimum cost of experiment validation efforts.

References

[1] Z. Li, S. Wang, W.S. Chin, L.E. Achenie, H. Xin, High-throughput screening of bimetallic catalysts enabled by machine learning, J. Mater. Chem. A Mater. Energy Sustain. (2017). doi:10.1039/C7TA01812F.

[2] Z. Li, X. Ma, H. Xin, Feature engineering of machine-learning chemisorption models for catalyst design, Catal. Today. 280 (2017) 232–238.

[3] X. Ma, Z. Li, L.E.K. Achenie, H. Xin, Machine-Learning-Augmented Chemisorption Model for CO2 Electroreduction Catalyst Screening, J. Phys. Chem. Lett. 6 (2015) 3528–3533.

531208 Prospects for Solving Micro-Kinetic Models with Automatic Differentiation and Regression

Abstract:

Micro-kinetic models provide a link between the molecular-scale mechanism of a catalytic reaction and the macro-scale observation of catalytic activity, and are central to computational catalysis. Micro-kinetic models commonly employ the mean-field approximation, which results in a series of coupled ordinary differential equations (ODE's). These ODE's are typically solved under simplifying assumptions such as a single rate-limiting step, or a quasi-steady state; the resulting equations are typically solved through standard numerical techniques such as finite differences or root finding. While these approaches provide valid solutions, each different approximation requires a different solution method. Similarly, coupling the micro-kinetic models to reactor models or including more complex rate constant expressions often requires substantial changes to the solution approach. Furthermore, the resulting numerical solutions are difficult to integrate with parameter fitting or uncertainty estimates. In this talk a different approach to solving micro-kinetic models will be introduced. This approach utilizes a combination of automatic differentiation and non-linear regression in order to establish a continuous function that approximates the solution to the micro-kinetic model. The approach is flexible, and the same basic framework can be used to solve the model with various types of approximations, including with constraints on parameter values; furthermore, the resulting solution is easily differentiable with respect to parameters, facilitating sensitivity analyses and parameter fitting. The approach is demonstrated for several model systems and strengths and weaknesses compared to traditional numerical techniques are discussed.

537707 Corresponding States Theory for Adsorption on Energetically Rough Surfaces

Abstract:

CORRESPONDING STATES THEORY FOR ADSORPTION ON ENERGETICALLY ROUGH SURFACES

Kaihang Shi, Erik Santiso and Keith E. Gubbins

Department of Chemical& Biomolecular Engineering, North Carolina State University,

Raleigh, NC 27695, U.S.A.

The Principle of Corresponding States was first proposed by van der Waals in 1881, based on his equation of state. The Principle made it possible to predict the vapor pressure curve, including the critical and normal boiling points, of liquids for which no measurements exist, based on experimental data for other liquids. It quickly found use in this way for predicting liquid properties for hydrogen and helium, neither of which had been liquefied at the time. When these light gases were later liquefied the predictions were found to be accurate. A more powerful derivation of the Principle is based on statistical thermodynamics, which makes clear the assumptions involved and provides a path to improve the theory so that it is more generally applicable. Subsequent generalizations of the theory of corresponding states, in particular that due to Pitzer, provide the basis for the prediction of many physical properties of homogeneous gases and liquids, and are widely used in process simulation. However, these correlations are not applicable to highly inhomogeneous or nanoscale systems, such as small drops, nanoparticles, adsorbed thin films on solid substrates, polymer brushes, etc. A statistical mechanical analysis of the properties of a highly inhomogeneous nanoscale system shows that the equilibrium properties (adsorption isotherms, isosteric heats, phase change conditions, etc.) are determined by a small number of dimensionless parameters, including the dimensionless width, H^*, of the nanophase and a *microscopic wetting parameter*, α_w, that is defined in terms of intermolecular forces and solid structure; α_w is a measure of wetting that applies at all scales and for any kind of adsorbed film (gas, liquid or solid) [1,2].

We illustrate the usefulness of this approach using examples drawn from both experimental and molecular simulation studies of gas-liquid and liquid-solid phase separations [3,4], and pressure enhancement in nanopores [5-7], with emphasis on simple pore geometries. These examples illustrate the central role played by wetting, and also the breakdown of some concepts and macroscopic laws, such as the equations of Kelvin, Laplace and Gibbs-Thomson, for nano-phases confined within small pores. They also suggest consistency tests for experimental data that should be applied in the event of capillary condensation or other phase changes in mesoporous materials.

The approach can be extended to account for geometric or energetic wall roughness [8], and to adsorbate molecules that interact with site-site potentials having both Lennard-Jones and point charge sites.

Finally, we describe an application of corresponding states theory (called *conformal sites theory*) to describe adsorption on solid surfaces that are energetically heterogeneous, i.e. have a variety of adsorption sites. The basic assumption in this theory is that all of the intermolecular interactions between adsorption sites on the surface and molecules in the adsorbed film conform to the same functional form, which includes dispersion, overlap and Coulomb interactions. The theory maps the heterogeneous surface onto a homogeneous one, in which all adsorption sites have the same energy. Through comparisons with molecular simulations for both the heterogeneous surface and the homogeneous reference surface, we show that the theory is accurate even when there is a large difference in energy between the various surface sites.

Figure 1. Experimental results for the effect of confinement in cylindrical pores on the melting point of various adsorbates. Here CNT = carbon nanotube and α_w is the microscopic wetting parameter; high values of α_w indicate strong wetting. For large pore widths the temperature shift is approximately linear, as required by the Gibbs-Thomson equation, but departures occur for smaller widths.

1. R .Radhakrishnan, K.E. Gubbins and M. Śliwinska-Bartkowiak, "Global Phase Diagrams for Freezing in Porous Media", *Journal of Chemical Physics*, **116**, 1147-1155 (2002).

2. Keith E. Gubbins, Yun Long and Malgorzata Sliwinska-Bartkowiak, "Thermodynamics of Confined Nano-Phases", *Journal of Chemical Thermodynamics*, **74**, 169-183 (2014).

3. L.D. Gelb, K.E. Gubbins, R. Radhakrishnan and M. Sliwinska-Bartkowiak, "Phase Separation in Confined Systems", *Reports on Progress in Physics*, **62**, 1573-1659 (1999).

4. C. Alba-Simionesco, B. Coasne, G. Dosseh, G. Dudziak, K.E. Gubbins, R. Radhakrishnan and M. Śliwinska-Bartkowiak, "Effects of Confinement on Freezing and Melting", *Journal of Physics: Condensed Matter*, **18**, R15-R68 (2006).

5. Yun Long, Jeremy C. Palmer, Benoit Coasne, Małgorzata Śliwinska-Bartkowiak and Keith E. Gubbins, "Pressure enhancement in carbon nanopores: A major confinement effect", *Physical Chemistry Chemical Physics*, **13**, 17163-17170 (2011).

6. H. Drozdowski, M. Kempinski, M. Śliwinska-Bartkowiak, M. Jazdzewska, Y. Long, J.C. Palmer and K.E. Gubbins, „Structural Analysis of the Behavior of Water Adsorbed in Activated Carbon Fibers", *Physical Chemistry Chemical Physics*, **14**, 7145-7153 (2012).

7. Yun Long, Jeremy C. Palmer, Benoit Coasne, Małgorzata Śliwinska-Bartkowiak, George Jackson, Erich A. Müller and Keith E. Gubbins, "On the Molecular Origin of High Pressure Effects in Nanoconfinement: Effects of Surface Chemistry and Roughness", *Journal of Chemical Physics*, **139**, 144701 (2013).

8. M. Śliwinska-Bartkowiak, A. Sterczyńska, Y. Long and K.E. Gubbins, "Influence of Microroughness on the Wetting Properties of Nano-Porous Silica Matrices", *Molecular Physics*, **112**, 2365-2371 (2014).

527085 Trade-Off between CO_2/CH_4 Membrane Permeability and Selectivity: Mesoscale Insights from Molecular Simulation

Abstract:

Trade-off between CO_2/CH_4 membrane permeability and selectivity: Mesoscale insights from molecular simulation

Xiaohua Lu

College of Chemistry and Chemical Engineering, State Key Laboratory of Materials-Oriented Chemical Engineering, Nanjing Tech University, Nanjing, 210009, PR China

The separation of CO_2/CH_4 is important in both biogas upgrading and natural gas sweetening. In the past two decades, gas membrane separation technology has become much more advanced due to interest in the technology¡äs reliability, low-energy consumption, and minimal capital investment. However, most membranes used in industry today are still limited due to an inherent upper bound limitation on their performance, showing a conventional trade-off behavior between permeability and selectivity [1].

It is believed that membrane process is a compromise in competition between two dominant mechanisms: the permeability mechanism and the selectivity mechanism[2]. In essence, the system behavior is dominated by the complex structure of the membranes and the complex interaction between the gas and the membranes. However, no quantitative results of the permeability and selectivity with the structure structural properties and the interaction were obtained, makes it different to revealed the inherent mechanism to break the Robeson upper bound.

In this work, a novel method based on linear nonequilibrium thermodynamics[3] and equilibrium molecular simulation was developed to *quantitatively* predict selectivity and permeability for CO_2/CH_4 mixtures based on the structural properties of materials such as effective contact surface area. By adjusting the potential parameters (¦Å, ¦Ò) artificially, different effective contact surface area was obtained and the optimal region to break the Robeson upper bound was obtained.

A new channel was formed by the combination of the CO_2 molecules with the carbon membranes, and the original CH_4 channel was changed. The CH_4 confinement space becomes smaller, which improves the selectivity and breaks through the Robeson limit. The formation of the new "limited domain" structure is mainly affected by the solubility of CO_2 and CH_4.

The research methods developed in this paper can be further extended to other systems, such as mass transfer in porous materials of N_2, H_2S, H_2O. It can also be expanded to other systems such as catalytic reaction system.

Keywords: molecular simulation, linearized non-equilibrium thermodynamics model, membrane separation

References:

[1] H. B. Park, J. Kamcev, L. M. Robeson, M. Elimelech, B. D. Freeman, ¡°Maximizing the right stuff: The trade-off between membrane permeability and selectivity¡±, Science, **2017**, 356, 1137.

[2] J. H. Li, ¡°Exploring the Logic and Landscape of the Knowledge System: Multilevel Structures, Each Multiscaled with Complexity at the Mesoscale¡±, Engineering, **2016**, 2, 276-285.

[3] W. L. Xie, X. Y. Ji, X. Feng, X. H. Lu, ¡°Mass-Transfer Rate Enhancement for CO_2 Separation by Ionic Liquids: Theoretical Study on the Mechanism¡±, AIChE Journal, **2015**, 61, 4437-4444.

539542 Isobaric Vapor-Liquid Equilibrium of Monoterpene + Sesquiterpene System at Normal Pressure

Abstract:

Isobaric vapor–liquid equilibrium (VLE) data for binary systems composed of 3-carene (monoterpene), *p*-cymene (monoterpene) and *β*-caryophyllene (sesquiterpene) were measured using a modified equilibrium still at 101.3 kPa. Both the Herrington method and Van Ness point-to-point method were adopted to confirm the thermodynamic consistency of the experimental VLE data, for which the test results showed good thermodynamic consistency. The experimental VLE data were correlated by Wilson, NRTL, and UNIQUAC models and the binary interaction parameters were obtained. The deviations of bubble-point temperature and vapor–phase composition between calculated values and experimental values were small, which meeted both the require of engineering design and application.

526479 Reaction Density Functional Theory and Its Application to Reactions in Aqueous Solution

Abstract:

Accurate determination of reaction barrier is crucial in the fields of chemical reaction engineering since itâ€™s directly associated with the reaction rate and reaction path. Whereas massive quantum calculations relying on commercial softwares have been reported for understanding reaction mechanisms, very little attention has been paid to the solvent effect on the reaction barrier. The difficulty originates partially from the ill capture of inhomogeneous solvent structure around the reaction site. Herein, we propose a reaction density functional theory (RxDFT) developed by combining quantum density functional theory for calculating intrinsic reaction energy with classical density functional theory for addressing solvation contribution. For demonstration, two chemical reactions in aqueous solution are revisited by using RxDFT. We illustrate that the free energy profiles of reaction barrier/path can be well predicted, in comparison with the results of QM/MM simulation. Since no solvent molecule is explicitly considered, the developed RxDFT is promising to provide an efficient alternative to QM/MM approach or *ab initio* calculation.

528231 Curvature Dependence of Henry's Law Constant and Non-Ideality of Gas Equilibrium for Highly Curved Vapor-Liquid Interfaces

Abstract:

Curved interfaces between coexisting vapor and liquid phases are ubiquitous in nature. The recently confirmed existence of surface/bulk nanobubbles raises the question of whether the law is applicable for highly curved vapor/liquid (VL) interfaces. By combining molecular dynamics simulations and thermodynamic analysis, we investigated the applicability of Henry's law for highly curved VL interfaces. In general, the Henry's law could not be used directly for nanobubbles that have highly curved vapor-liquid interfaces. Instead, the Henry's law constant exhibits a novel curvature dependence, which is ascribed to the non-ideality of gas in liquid phase because of the required gas supersaturation for stabilizing the nanobubbles. To account for this effect, we developed a method for the determination of the Henry's law constant from the level of gas supersaturation. With the modifed Henry's law, we trun to vapor/liquid equilibrium and gas partitioning across highly curved VL interfaces. Again, the non-ideality of gas in liquid significantly changes the relationship between gas supersaturation and nanobubble radius. Our findings can provide guidance for determining the gas partitioning of nanobubbles and explaining its unusual phase equilibrium behaviors.

531908 Nanostructured Photoelectrodes for Enhanced Charge Carrier and Mass Transfer in Solar Water Spitting

Abstract:

Photoelectrochemical water splitting for hydrogen production by semiconductors has been extensively investigated in recent decades for the conversion and storage of the intermittent solar energy. During this process, photo-generated charge carriers will transfer from the bulk to the surface of photoelectrodes for the water splitting reaction. Meanwhile, hydrogen and oxygen bubbles formed upon the surface reaction will detach from the surface of photoelectrodes. However, unfavorable surface and interface structures will hinder the transfer of charge carriers, which severely limits the overall efficiency. Additionally, a poor wettability of the electrode will result in the accumulation of hydrogen and oxygen bubbles, which imposes a significant mass transfer resistance between electrolyte and electrode.

This research describes the design and synthesis of nanostructured photoelectrodes for enhanced charge carrier transfer and bubble detachment. Fe_2O_3 nanotubes was synthesized with shortened distance for charge collection, exhibiting a photocurrent density of 1.5 mA/cm2 at 1.23 V vs. RHE. A dendritic nanostructured Fe2O3 photoanode was also obtained for enhanced solid-liquid junction area to improve the charge collection efficiency, generating a photocurrent of ~2.5 mA cm^{-2} at 1.23 V vs. RHE. For Si based photoanodes that suffer from interface defects, a high-quality Al_2O_3 tunnelling layer was introduced to repair the interfacial defects and suppress charge recombination, yielding a photocurrent density of ~ 28 mA cm^{-2} at 1.23 V vs. RHE for oxygen evolution. Eventually, planar Si(100) substrate was etch to expose the (111) facets, forming micro-pyramid structures. The enhanced wettability of the pyramid structured Si photocathode exhibited a water reduction photocurrent of 35 mA cm^{-2} at 0 V vs. RHE, which was caused by the promoted detachment of hydrogen bubbles.

528008 Dynamic Interfacial Tension during Droplet/Bubble Generation at Microchannel Junctions - an Important Mesoscale Scientific Issue in Multiphase Microflow Study

Abstract:

The dynamic interfacial tensions caused by surfactant adsorption and chemical reactions during droplet/bubble generation process at microchannel junctions are important mesoscale scientific issue between molecule and microchannel scales. This phenomenon exhibits the variations of interfacial composition during fluid particle generations with chemical processes, which is commonly taken place in microflow reaction and separations. This presentation gives a summary of our recent studies focusing on the determination of dynamic interfacial tension at during the droplet and bubble generations T-shaped[1] and coflowing microchannel junctions[2], and the variation laws of interfacial compositions with several surfactant absorption processes[3] as well as the chemical reaction of CO_2 and MDEA solution[4]. Experimental results show increase of decrease in interfacial tension from the static interfacial tension with equilibrium states. The Marangoni effect caused the gradient of dynamic interfacial tension is further exhibited by LBM simulations[5], which is helpful for the understanding of multiphase microflow processes. Funding supports are from National Natural Science Foundation of China (21776150, 91334201, U1302271).

References:

1. Wang K, Lu YC, Xu JH, Luo GS. Determination of dynamic interfacial tension and its effect on droplet formation in the T-shaped microdispersion process. Langmuir. 2009; 25: 2153-2158.
2. Wang K, Xie LS, Lu YC, Luo GS. Generating microbubbles in a co-flowing microfluidic device. Chem Eng Sci. 2013; 100: 486-495.
2. Wang K, Zhang LM, Zhang WL, Luo GS. Mass-transfer-controlled dynamic interfacial tension in microfluidic emulsification processes. Langmuir. 2016; 32: 3174-3185.
4. Yang L, Liu G, Luo S, Wang K, Luo G. Investigation of dynamic surface tension in gas-liquid absorption using a microflow interfacial tensiometer. React Chem Eng. 2017; 2: 232-238.
5. Riaud A, Zhang H, Wang XY, Wang K, Luo GS. Numerical study of surfactant dynamics during emulsification in a T-junction microchannel. Langmuir. 2018; doi: 10.1021/acs.langmuir.8b00123.

538236 Modeling of Interfacial Behaviors of Isobutane Alkylation with 2-Butene Catalyzed By Sulfuric Acid/BrÃ¸nsted Acidic Ionic Liquid

Abstract:

Modeling of Interfacial Behaviors of Isobutane Alkylation with 2-butene Catalyzed by Sulfuric Acid/Brønsted Acidic Ionic Liquid

Weizhong Zheng[1], Piao Cao[1], Weizhen Sun[1], and Ling Zhao[1]

[1] *State Key Laboratory of Chemical Engineering, East China University of Science and Technology, 200237, Shanghai, China, wzzheng728@foxmail.com, 1621756797@qq.com, sunwz@ecust.edu.cn, zhaoling@ecust.edu.cn*

The alkylate produced by the alkylation of isobutane with C4 olefins is a desirable blending component for high quality gasoline.[1] Although the alkylation processes for commercial importance use either sulfuric or hydrofluoric acid as catalyst, both of them suffer from fatal drawbacks.[2] In recent years, ionic liquids (ILs) have drawn considerable attentions for their applications in C4 alkylation owing to their chemical stability, negligibly low vapor pressure, and Brønsted and Lewis acidity.[3] As a dual solvent-catalyst, Brønsted acidic ionic liquids (BILs) can enhance obviously the catalytic activities of sulfuric acid, when used as a co-catalyst with the sulfuric acid. The sulfonic acid functionalized BILs (SFILs) and non-SFILs not only can improve the catalytic activities, but also can enhance greatly the stability of the sulfuric acid.[4] Actually, the BILs act as a surfactant which can improve the acid/hydrocarbon interfacial properties and further enhance the catalytic activities of sulfuric acid, because C4 alkylation catalyzed by sulfuric acid is a typical liquid/liquid interfacial reaction. However, it is difficult to detect the interfacial properties using experimental methods. Alternatively, the molecular dynamics simulation is proved to be a proper tool to probe the interfacial behaviors involving the C4 alkylation.

The interfacial properties between C4 hydrocarbons and the SFILs with different alkyl chain length were modeled using molecular dynamic simulation. Compared to the pure H_2SO_4 systems, BILs can enhance the dissolution and diffusion of isobutane at interface, as shown in Figure 1. The SFILs with longer alkyl chain show stronger density enrichment at interface and tend to protrude more deeply into the reactant phase with their orientation perpendicular to the interface, which is more beneficial to the dissolution of isobutane. Compared to the non-SFILs, the SFILs can facilitate a higher dissolution of isobutane but inhibit its diffusion at interface. The good agreement between simulations and experiments suggests that the solubility and diffusion of isobutane play a synergized role in the quality of the alkylate. The fundemental understanding in this work can provide a novel approach to screen and design SFILs for isobutane alkylation.

Figure 1. Equilibrated snapshots of MD simulation. (a) H_2SO_4, (b) [PMim][HSO_4], (c) [BMim][HSO_4], (d) [HMim][HSO_4], (e) [OMim][HSO_4], (f) [PSMim][HSO_4], (g) [HSMim][HSO_4], (h) [OSMim][HSO_4]. The sulfuric acid in the middle of the box is shown in the right side.

The financial support by the National Natural Science Foundation of China (91434108) is gratefully acknowledged.

[1] Welton T. Room-temperature ionic liquids. Solvents for synthesis and catalysis[J]. Chemical reviews, 1999, 99(8): 2071-2084.

[2] Busca G. Acid catalysts in industrial hydrocarbon chemistry[J]. Chemical reviews, 2007, 107(11): 5366-5410.

[3] Corma A, Mart"ªnez A. Chemistry, catalysts, and processes for isoparaffin-olefin alkylation: actual situation and future trends[J]. Catalysis Reviews-Science and Engineering, 1993, 35(4): 483-570.

[4] Amarasekara A S. Acidic ionic liquids[J]. Chemical reviews, 2016, 116(10): 6133-6183.

531971 Computational Modeling of Thermal Diffusion at the Mesoscale

Abstract:

Heat transfer in crystalline solids occurs through the motion of packets of vibration waves, or phonons. Over sufficiently large length scales, the motion of the phonons is diffusive, local thermal equilibrium is maintained, and the relation between the heat flux and temperature gradient is accurately described by Fourierâ€™s heat law. However, at nanometer length scales or the mesoscale where characteristic lengths are comparable to the phonon mean-free path, phonon transport becomes increasingly ballistic and Fourierâ€™s law is therefore inaccurate. Applications requiring mesoscale simulations of heat transfer include microelectronics devices and improved models for hot spot formation and growth in energetic materials. The phonon Boltzmann transport equation (BTE) is well-suited to these types of problems because it captures behavior over the entire ballistic-diffusive spectrum. In the present work, we assess the feasibility of using molecular dynamics (MD) simulations to obtain the material properties needed to solve the BTE for graphene. We then simulate a graphene sheet subject to different thermal conditions using the BTE and compare the resulting temperature profiles with analogous results obtained by MD. Prepared in part by LLNL under Contract DE-AC52-07NA27344.

529575 Translocation Energy Calculation on Human β Defensin Type 3 through Bacterial Lipid Membranes

Abstract:

Human β Defensin type 3 (hBD-3), mainly secreted by the human epithelial cells, is a cysteine-rich small peptide, which has 45 residues and a charge density of +11. It can form a dimer or higher order oligomer. It has antimicrobial activity even at high salt concentrations, of which hBD-3 is believed to interact with the bacterial membrane, breaking the membrane, causing cell leakage, thus killing the bacterial cell. To better understand the anti-bacterial mechanism of hBD-3, we have calculated free energies of hBD-3 monomer, dimer and tetramer through two different types of bacterial lipid bilayers. One kind is represented by 1-Palmitoyl-2-oleoyl-sn-glycero-3–phosphatidylserine (POPS) bilayer, and the other is the mixture of POPS with 1-Palmitoyl-2-oleoyl-sn-glycero-3 –phosphatidylcholine (POPC), to consider the lipid membrane chemical composition effect. Besides that, hBD-3 dimer through POPC bilayer was also performed to as a comparison, with POPC bilayer resembling neutrally charged normal cell membrane. Umbrella sampling simulations along with coarse-grained molecular dynamics were performed to calculate the potential of mean force (PMF) associated with hBD-3 monomer and oligomers translocating through different kinds of membranes. Based on PMF results up to now, hBD-3 dimer cannot trans-pass the neutrally charged lipid membrane, while only needs to overcome a much lower energy barrier to transpass the negatively charged lipid membrane. hBD-3 dimer needs to encounter a similar free energy barrier to hBD-3 monomer to cross the POPS lipid bilayer. Besides that, due to the electrostatic interaction between positively charged hBD-3 and the negatively charged head of POPS lipid, hBD-3 inserting into the negatively charged membrane by no more than 16.2 Å is energetically favorable.

532840 Efficient Discovery of Novel Molecules: How to Uncover Gems in the Haystack

Abstract:

A fundamental challenge in molecular science and engineering is uncovering target molecular structures with optimal properties. This inverse design question underlies almost all of chemistry and chemical engineering -- how to find the molecules for new antibiotics and pharmaceuticals, to molecular catalysts and materials design for energy conversion and storage, such as photovoltaics, piezoelectrics, spintronics, and many others. Computational discovery, particularly the use of quantum chemical methods and density functional approximations have expanded the accuracy of property predictions. We will outline the use of open source chemistry software developed in our group, including Avogadro - an intuitive molecular editor, Open Babel - a cheminformatics toolkit, and Lamark - an evolutionary algorithm for rapidly finding target molecules with optimal properties. Particular focus will be on organic photovoltaics, molecular piezoelectrics, and dielectrics, with discussion of integrating multiple programs into scalable workflows, and rapid data analysis and visualization.

538365 Chemlg - a Smart and Massively Parallel Code to Accelerate the Molecular Library Generation

Abstract:

The discovery of new compounds, materials, and chemical reactions with exceptional properties is the key to progress in chemistry. This process can be dramatically accelerated by means of the virtual high-throughput screening of large-scale candidate libraries. The key challenge is that chemical space is practically infinite, and any approach to survey it or enumerate certain of its domains has to address the problem of combinatorial complexity. A number of software packages are now available for the enumeration of compound space. However, the majority of these packages deal with small molecules, which are relevant as drug-like molecules. By far the largest small molecule database, GDB-17, contains 166.4 billion molecules, which are combinatorically generated with up to 17 atoms consisting of C, N, O, S, and halogens. These screening libraries have great utility in drug-related research, but larger molecules are required for various other applications. However, as mentioned before, an exhaustive list of large molecules would result in a combinatorial explosion of candidates, and would thus be impractical for screening.

Our work aims to extend and generalize library generation to identify molecular lead candidates and reaction networks in various other applications such as functional polymers, optoelectronics, and catalysis. Our massively parallel generator ChemLG is part of our ChemHTPS program suite for automated, virtual high-throughput screening studies, and it offers a multitude of options to customize and restrict the scope of the enumerated chemical space and thus tailor it for the demands of specific applications. To streamline the non-combinatorial exploration of chemical space, we incorporate genetic algorithms into the framework. Genetic algorithms have shown to be effective in optimizing chemical structures and generating useful compounds for different target applications. We built the code in python and implement parallelization using mpi4py library. In addition to implementing smarter algorithms, we also focus on the ease of use, workflow, and code integration to make this technology more accessible to the community.

532718 GOMC: GPU Optimized Monte Carlo for the Simulation of Phase Equilibria and Physical Properties of Complex Fluids.

Abstract:

GPU Optimized Monte Carlo (GOMC) is open-source software for simulating many-body molecular systems using the Metropolis Monte Carlo algorithm. It supports simulations in a variety of ensembles, which include canonical, isothermal-isobaric, grand canonical, and Gibbs ensemble. This allows GOMC to be used to study vapor-liquid and liquid-liquid equilibria, adsorption in porous materials, surfactant self-assembly, and condensed phase structure for complex molecules. GOMC supports a variety of all-atom, united atom, and coarse grained force fields such as OPLS, TraPPE, Mie, and Martini. The software has been written in object oriented C++, and uses OpenMP and NVIDIA CUDA to allow for execution on multi-core CPU and GPU architectures. The combined multi-core CPU and GPU parallelization achieves up to two orders of magnitude speed-up compared to serial execution.

GOMC is available from GitHub (https://github.com/GOMC-WSU).

538430 Lancelot - an Open-Source Codebase for Simplified Computational Chemistry

Abstract:

Computational chemistry research is a vital tool for the study of atomic and molecular reactions and mechanisms. Unfortunately, many tedious tasks exist in the day-to-day lives of computationalists, such as parsing common file formats, interfacing with software, and automating job submissions to queueing systems on servers. Lancelot is a code base, developed at Cornell to improve work flow by removing these, and other, simple tasks. Written in python, with extensive documentation, it is quick to pick up and install. Hooks are written to allow for automated Molecular Dynamics simulations (using LAMMPs) and Density Functional Theory simulations (using Orca and Gaussian09). Further, it includes an extensible Nudged Elastic Band package that interfaces with Orca and Gaussian09 by default, allowing for the capture of Minimum Energy Pathways (MEPs). This method is written in an object-oriented approach, so as to allow for an extension to any software. Finally, Lancelot also incorporates some automated post-processing of DFT simulations using the Visual Molecular Dynamics (VMD) software, developed at UIUC. Lancelot is available online for any interested users, and is currently licensed under the CC-BY-NC 3.0 license.

538338 Saffire: Enabling Large Scale Simulations of Rare Events

Abstract:

We have developed software platform SAFFIRE to perform large scale forward flux sampling (FFS) simulations. FFS is an established scientific method for sampling rare events in simulations. However, as the difficulty of the scientific problem increases, the amount of data and the number of tasks required for FFS is challenging to manage with traditional scripting tools and languages for high performance computing. SAFFIRE software framework has been developed to address these challenges. SAFFIRE utilizes Hadoop to manage a large number of tasks and data for large-scale FFS simulations. The framework is shown to be highly scalable and is able to the support large-scale FFS simulations necessary to extend the use of FFS to complex molecular systems on commodity cluster computing systems. In the talk we will discuss the journey of developing the software highlighting effective and ineffective strategies, as well as discuss plans for sustaining the software and training students and other groups to use this platform.

538580 Quantifying Nanostructure within Molecular Simulations Using Geometry-Based Criteria

Abstract:

Molecular simulations have the potential to track the detailed positions and orientations of molecules over time or within a statistical mechanical ensemble. That wealth of information creates a corresponding challenge of inferring well-sampled physics from the underlying correlations and dynamics, particularly within multicomponent systems in which each molecule brings its own level of complexity. Our group has been revamping geometric analysis methods based on Delaunay and Voronoi tessellation so they yield measures of local microstructure within systems of heterogeneous composition. These tools aid with interpreting intermolecular packing and organization beyond the extent that can be attained using a radial distribution function. The tessellation process divides a simulation cell into a unique set of tetrahedra and polyhedra that fill the box. The shape and the physical extent of a microstructure domain are identified by neighboring polyhedra that share common features. In multicomponent simulations that represent a bitumen-water system, as an example, connected tetrahedra that are each defined by locations of 4, 3, or even 2 atoms of water molecules map the location of a water droplet. The geometric analysis method allows the average droplet volume and spatial extent to be sampled along a simulation trajectory. Tetrahedra with 2 atoms each from water and from bitumen define the interface, enabling direct calculation of its fluctuating shape. In a simulation of a bitumen that is modified by a "biobinder", the geometric analysis quantifies domains of different chemical classes within the bulk system as well as the partial molar volume of each molecule type in a highly multicomponent system. Existing capabilities to identify nanoscale voids within molecular simulation systems are being used to quantify sorption sites and diffusion paths within modified bitumen systems

535502 Runtime Code Generation for User-Configurable Metropolis Monte Carlo Energy Evaluation in HOOMD-Blue

Abstract:

We have developed a module for HOOMD-blue for that enables parallel Metropolis Monte Carlo (MC) simulations of hard particle cores with user-defined pairwise energetic interactions. Previously, HOOMD-blueâ€™s Monte Carlo capabilities were limited to hard particle interactions alone. The new module uses the low level virtual machine (LLVM) and the open-source Clang compiler to compile user-provided C++ code down to efficient machine code at run time. Users can express any pairwise interaction potential that is a function of the relative particle positions, and the orientations, types, diameters, and charges of the two particles. The code is compiled at runtime, so new models can quickly be developed and tested with no need to write multiple source files, user interfaces, documentation, or recompile and install a plugin each time a change is made. The user-provided energy evaluation is evaluated in HOOMD-blueâ€™s existing optimized parallel MC engine, enabling high performance simulation execution. We also implemented the geometric cluster algorithm of J. Liu and E. Luijten [1], and an efficient code path for evaluating the pairwise energy between two rigid bodies built of many interaction sites with user-defined energies evaluated between all pairs of interaction sites within a cutoff. These capabilities are available in HOOMD-blue open source v2.3+.

[1]: J. Liu and E. Luijten, Phys. Rev. Lett. **92**, 035504 (2004).

532313 Implementation of Harmonically Mapped Averaging Methods in Popular Molecular Simulation Codebases

Abstract:

"Mapped averaging" is a recently published scheme for the reformulation of ensemble averages. The framework uses approximate results from statistical mechanical theory to derive new ensemble averages (mapped averages) that represent exactly the error in the theory. Well-conceived mapped averages can be computed by molecular simulation with remarkable precision and efficiency, and in favorable cases the speedup factors are several orders of magnitude. For crystalline systems, mapped averaging enables simulation to compute directly the anharmonic contribution to the properties, without noise contributed by harmonic behavior. The result is a technique for computing crystalline properties with unprecedented, transformative efficiency. The aim of this project is to implement these methods on well-established and widely used software packages for simulation of crystalline systems, and furthermore to develop mapped averages for new applications of interest to the users of these systems. Specifically, we are developing implementations for LAMMPS, HOOMD, Cassandra, and VASP, which altogether have a base encompassing thousands of users. Software elements implemented in this project are in many cases completely transparent to the users of these packages, and can be employed by them with no added complication, to speed up their calculations by orders of magnitude.

We describe our progress in this development and demonstrate mapped averaging methods on some of the targeted platforms.

533449 Catkit: Symmetry Methods for Automated Generation of Catalytic Structures

Abstract:

Cheminformatics have dramatically accelerated the rate at which we discover useful properties of chemical species, originally for the purpose of drug discovery. This acceleration is made possible by taking advantage of bond symmetries present in organic molecules, allowing computers to systematically enumerate through an otherwise infinite search space. These techniques are also easily extend to other forms of symmetry, such as the space group symmetries present in bulk systems.

Since the systems in heterogeneous catalysis contain the complexity of both of these chemical domains, high-throughput techniques remain difficult to automate effectively for catalyst design. In this work, we propose a methodology which combines space group and bond symmetry techniques general enough to enumerate a large number of possible catalytic systems. This includes surfaces of any miller index produced from a provided bulk system, unique adsorption site identification, and creation of 3D structures bonded by one or multiple sites. This technique also produces a graph structure which provides the basis for all machine-learning fingerprints which require connectivity information; ideal for accurate descriptors which are not dependent on information gathered by expensive techniques such as Density Functional Theory.

We demonstrate these techniques for the enumeration of unique adsorption configurations across a wide range of alloy systems and catalytically relevant adsorbates. Utilizing these enumerated structures we show how the graph of each can be easily utilized to construct a variety of fingerprints as a basis for machine-learning accelerated predictions of adsorption energies. Finally, we discuss our Python implementation of these algorithms in the open source software package known as CatKit, intended to make high-throughput screening techniques easier to implement for the community.

530615 Predicting Adsorption Properties on Bimetallic Alloys As a Function of Local Morphology and Atomic Composition

Abstract:

Bimetallic nanoparticles are widely used in catalytic and energy applications because they present a tunable configurational space together with synergistic effects between component metals which influences adsorption energies. These energies function as thermochemical reactivity descriptors in DFT derived screening models which have successfully designed bimetallic catalysts for numerous catalytic applications [1]-[5]. While current paradigms correlate surface adsorption to reactivity trends through a forward design approach, they lack a site-by-site resolution in descriptor space, hindering their capability of predicting active sites with atomic level precision. Incorporating such specificity enhances flexibility in materials space through targeted engineering of active sites, while simultaneously explicitly accounting for surface segregation phenomena that affects the stability of bimetallic catalysts.

Employing adsorption site stability as a descriptor, we predict adsorption energies of catalytically relevant descriptors like OH*, CH*, CH_3* and CO* on generic bimetallic alloys of Ag, Au, Cu, Ir, Pd, Pt, and Rh at a site-by-site resolution, across a broad morphological and compositional design space. We build this framework by extending recently postulated scaling relations between metal atoms and metal-adsorbate complexes for monometallic systems [6] to bimetallic alloys. We observe linear correlations between the inherent stability of a surface binding site and the adsorption energies of metal-adsorbate complexes. These correlations are determined from a limited set of slab-based DFT calculations, for fixed binding site compositions, while varying the chemical ordering of nearest-neighbor metal atoms. We directly estimate molecular adsorption energies from the scaling trend, while systematically quantifying the effects of modifications in morphology and chemical ordering on adsorption strengths. Slopes and intercepts of the scaling lines are discussed in the context of the d-band theory. We examine the versatility of our model by predicting adsorption energies of OH* on top and bridge sites of 147 atom cuboctahedral PtAu nano-alloys in varying coordination and compositional environments. When integrated with a coordination-based model that directly estimates adsorption site stability [7], scaling relations predict adsorption energies within 0.12 eV, validating our paradigm. Finally. we discuss potential applications of our approach towards nano-engineering active sites with atomic resolution, to discover the next generation of bimetallic catalysts.

References

[1] Studt F., Abild-Pedersen F., Bligaard T., Sørensen R. Z., Christensen C. H., Nørskov J. K., *Science* **2008,** *320,* 1230.

[2] Greeley J., Stephens I. E. L., Bondarenko A. S., Johansson T. P., Hansen H. A., Jaramillo T. F., Rossmeisl J., Chorkendorff I., Nørskov J. K. *Nat. Chem.* **2009,** 1, 552.

[3] Studt F., Sharafutdinov I., Abild-Pedersen F., Elkjær C. F., Hummelshøj J. S., Dahl S., Chorkendorff I. B., Nørskov J. K. *Nat. Chem.* **2014,** *6,* 320.

[4] Ma X., Li Z., Achenie L. E. K., Xin H., J. *Phys. Chem. Lett.* **2015,** *6,* 3528.

[5] Saravanan K., Kitchin J. K., von Lillenfeld O. A., Keith J. A., *J. Phys. Chem. Lett.* **2017,** *8,* 5002.

[6] Roling L. T., Abild-Pedersen, F., *Chem. Cat. Chem.* **2018,** doi:10.1002/cctc.201701841

[7] Roling L. T., Lin L., Abild-Pedersen, F., *J. Phys. Chem. C*, **2017,** *121,* 23002.

536838 Insights from Machine Learning on a Large Database of Adsorption Energies

Abstract:

Adsorption energies of important intermediates are often excellent descriptors for catalytic performance. This has spurred significant interest in efficient predictions of adsorption energies, using both physical and data scientific approaches. However, larger data sets are needed in order to ensure the accuracy and generality of the approach. Here, we use a database with over 5,000 adsorption energies of H on stepped Ag alloy surfaces to develop an effective machine learning model to predict adsorption energies. We use an iterative process where we begin with a simple model, examine the errors, and then improve the model based on the errors. This brings physical insight into adsorption along with an accurate model. In particular, we discover counterintuitive trends in how adsorption energies depend on the H atom's first and second neighbor shells. The final model, an extra forest based on only 25 features, provides a high accuracy with a root-mean-square-error of 0.05 eV.

528022 Methods to Exploit Large Datasets in Catalysis

Abstract:

Computational catalysis techniques, such as density functional theory (DFT), have historically generated datasets of limited size due to the relatively high cost of performing DFT calculations. The cost of gathering DFT data has been decreasing because of the proliferation of public databases and advances in both computing power and workflow management. The decrease in computational cost has produced larger datasets, which offers new opportunities and challenges for data analysis. We present various methods that we use to analyze larger datasets to successfully gain new insights into theoretical catalysis.

530936 Generalized Geometric Descriptors for Oxygen Reduction Activity on Transition Metal Sulfides

Abstract:

Sabatier-type activity volcano plots have been used to determine optimal adsorption characteristics for maximizing catalytic activity. In this process of catalyst design, an important step involves designing catalysts involves identifying the active sites for optimal adsorption strength and engineering catalyst materials with target active sites since it has been shown that the variation in binding energies across heterogeneous surface sites is strongly dependent on the local environment[1]. Therefore, catalyst discovery can be accelerated through structure-energy-activity relationships, which allow us to formulate the active site design as merely finding the optimal local coordination environment that maximizes activity. Recently, structure-activity relationships have been developed for metals[1,2], and we extend this approach to obtain a structure-energy-activity relationship for transition metal sulfides. Transition metal sulfides represent an attractive class for ORR[3] and their rich phase diagram allows tunability of activity. We employ density functional theory (DFT) to study ORR activity on various transition metal sulfides including scandium, titanium, vanadium, chromium, ruthenium, nickel, cobalt, molybdenum, manganese, copper and iron sulfides.

We illustrate the robustness of the developed geometric descriptor for an example class of nickel sulfides[4]. As observed on metals and metal oxides, we show that there exists scaling between oxygen intermediates, that leads to a volcano relationship for the limiting potential in the OH* adsorption energy. We identify a simplest geometric descriptor using a least-squares approach to obtain the structure-energy descriptor: , which has excellent predictability (RMSE = 0.16 eV). Therefore, to a first approximation, a simple counting of neighboring atoms around the surface site can be used to screen catalysts. Our analysis suggests that having 3 nearest-shell sulfur atoms leads to high activity for Ni-based sulfides.

We will discuss a generalization of this structure-energy descriptors for transition metal sulfides and the trends in the coefficients will be discussed to formulate a generalized structure-activity descriptor for the transition-metal sulfides class. These coefficients are expected to strongly depend on metal-metal and sulfur-metal bonding characteristics. Identification of structure-activity relationships for classes of compounds allows for rapid identification of candidates and robust activity quantification. This approach could allow for rapid screening within different material classes for a variety of reactions.

References:

1. Calle-Vallejo, F. et al. *Science*, 350 185 (2015).
2. Friebel, D., **Viswanathan, V**. et al. *Am. Chem. Soc.* 134 9664 (2012).
3. Behret, H. et al. G. *Acta,* 20 111 (1975).
4. Yan, B., **Krishnamurthy, D.,** Hendon, C., Deshpande, S., Surendranath, Y., **Viswanathan, V**, *Joule* 1, 600 (2017).

529093 Quantifying Confidence in DFT Predicted Surface Pourbaix Diagrams at Solid-Liquid Interfaces on Transition Metal Surfaces

Abstract:

Emerging electrochemical energy conversion and storage will play an increasingly crucial part in sustainable energy systems. Low temperature Proton-Exchange Membrane (PEM) fuel cells are attracting considerable attention as a means to convert chemical energy conversion. However, it is widely recognized that a major bottleneck for PEM fuel cells is related to sluggish kinetics of the Oxygen Reduction Reaction (ORR) at the cathode. Understanding how to improve ORR and the activity of catalyst materials requires a detailed understanding of atomistic surface dynamics.

Density Functional Theory (DFT) calculations have been widely used to identify active catalysts by constructing free energy diagrams incorporating the electrochemically stable surface structure [1]. We construct surface Pourbaix diagrams to capture the most stable state of the surface under reaction conditions of electrode potential and pH. A unique aspect of this work is the incorporation of error estimation techniques from within the Bayesian Error Estimation Functional with Van der Waals corrections (BEEF-vdW) exchange correlation function [2]. We developed a systematic approach to propagate the uncertainty associated with the energetics for the construction of surface Pourbaix diagrams [3]. Within this methodology, we assign a prediction confidence value (c-value) [4] to adsorption energies of intermediates governing the ORR. As a result, surface phase transitions are no longer sharp and now have a finite width originating from DFT uncertainty.

We illustrate the surface phase stability of varying quantities of adsorbed O* and OH* intermediates relevant to ORR. In particular, we discuss the phase transition boundary from OH* to O* on commonly studied transition metals Pt, Pd, Ir, Rh, and Ru while noting the adsorption characteristics for favorable ORR activity in terms of our prediction confidence. We compare against linear sweep cyclic voltammetry experiments and DFT-based results and observe the most predominant species are in good agreement with our predictions. The OH* descriptor has been shown to be the optimal descriptor for the four electron ORR because it has the lowest uncertainty associated with prediction [5]. Therefore, we report our comparison of transition metal surfaces in terms of their binding strength of OH* as Ru>Rh>Ir>Pd>Pt, where Ru binds OH* the strongest which is in agreement with prior DFT studies [6]. We believe this method presents the opportunity to robustly incorporate multiple surface phases in determining electrochemical reaction mechanisms and thereby accurately determining catalytic activity.

References:

[1] Hansen, H. A., Rossmeisl, J., NÃ¸rskov, J. K. Phys. Chem. Chem. Phys. 10, 3722â€"3730 (2008).

[2] Wellendorff, J., Lundgaard, K. T., MÃ¸gelhÃ¸j, A., Petzold, V., Landis, D. D., NÃ¸rskov, J. K., Bligaard, T., Jacobsen, K. W. Phys. Rev. B 85, 235149 (2012).

[3] Vinogradova, O., Krishnamurthy, D., Pande, V., Viswanathan, V. arXiv:1710.08407 [cond-mat.mtrl-sci]

[4] Houchins, G., Viswanathan, V. Phys. Rev. 96, 134426 (2017).

[5] Krishnamurthy, D., Sumaria, V., Viswanathan, V. J. Phys. Chem. Lett. 9, 3, 588-595 (2018).

[6] Karlberg, G. S. Phys. Rev. B 74, 153414 (2006).

528583 Accelerating Inorganic Discovery with Machine Learning and Automation

Abstract:

Computation has emerged as a powerful tool for accelerating the discovery of new materials and molecules: first through first-principles simulation in high throughput screening and very recently even further with machine learning (ML). Unique challenges remain for the accelerated discovery and design of inorganic complexes, despite the highly tunable electronic structure properties that make these materials so compelling for applications in energy storage and catalysis. Density functional theory (DFT) is the method of choice in most computational screening, but DFT is simultaneously too computationally costly and inaccurate for the predictive design of open-shell transition metal complex properties. I will describe our recent efforts to overcome both of these challenges through developing ML models (e.g., artificial neural networks) that predict key first-principles energetic (e.g., spin-state ordering, redox potential, and binding energies) and geometric properties to within the accuracy of the underlying simulation method that provides the training data. I will describe our efforts to use these ML models for extrapolative applications (i.e., design) by incorporating heuristics for ML model uncertainty in a genetic algorithm to discover new inorganic compounds. I will also describe how we are using ML models to overcome increased variability and limited applicability of linear free energy relationships (LFERs), which are widely employed in the solid state, between energies of key intermediates or transition states in a catalytic cycle for single-site and molecular catalysis. Our work on representative catalytic reaction steps in oxidation catalysis demonstrates how our strategies first demonstrated on equilibrium properties such as spin state ordering can be extended to catalysis as well.

538738 The Living Journal of Computational Molecular Science: Improving Community Use of Molecular Simulation Methods through a New Publishing Model

Abstract:

The Living Journal of Computational Molecular Science (LiveCoMS) provides an open source, peer-reviewed home for manuscripts which share best practices in molecular modeling and simulation. LiveCOMS is ultra-low cost, non-profit, and aims to incentivize, through publication, uncommon or unique article types that will improve standards of use for molecular simulation and modeling. These works are living articles, regularly updated on GitHub with community input, and can include perpetually updated reviews, articles describing best practices, tutorials, comparisons between software packages, and other articles which aim to improve the use of simulation by practitioners in the field and also benefit from ongoing updates.

LiveCOMS is now open for submission, with planned first issue in Q4 of 2018. I will discuss how LiveCOMS meets needs, both for authors and for readers, that other journals do not, and the benefits of publishing an article in LiveCOMS. I will also describe how the process of updating a journal article, unique to LiveCOMS, can substantially increase the utility of the material in the article over the long term.

528122 Strategies and Software for Accelerating Inorganic Molecular Design

Abstract:

Chemical space is vast, with best estimates suggesting we have as yet characterized a tiny fraction of all possible compounds. The need for efficient discovery of new inorganic molecules as building blocks for materials and catalysts mandates that we identify smart ways to map out and explore transition metal chemical space. Virtual high throughput screening, typically driven by first-principles, density functional theory (DFT) calculations, has emerged as a powerful tool for the discovery of new materials. However, approximate DFT is simultaneously too computationally costly for full enumeration of chemical space and also too inaccurate for robust prediction of properties with correlated electrons. I will describe our recent development of the first open source toolkit for inorganic molecular discovery, molSimplify, that aims to address both these challenges. i) We introduce a divide-and-conquer approach to generate precise geometries of new inorganic complexes using building blocks from multi-million molecule organic libraries for automated simulation of diverse compounds with DFT. ii) We train and incorporate a neural network to predict quantum mechanical properties in microseconds for use in efficient screening workflows as opposed to minutes or hours with DFT. iii) We develop automated strategies to uncover structure-property correlations for molecular design. iv) We reveal structure-method-sensitivity relationships that guide when going beyond approximate-DFT is essential. v) We incorporate all these advances into evolutionary algorithm workflows in the molSimplify automated design (mAD) toolkit that adapt on the fly from data-driven models to DFT and beyond for robust prediction. I will review how implementing all of these techniques in our open source toolkit has advanced applications in catalysis and materials.

535153 Computational Materials Education and Training (CoMET): A Graduate Training Program to Teach Fundamentals of Quantum Density Functional Theory to Engineers and Scientists

Abstract:

Density functional theory will play a significant role in the discovery of new materials that offer solutions to the most pressing technological problems facing our society. Yet no engineering curriculum exists at many US universities beyond basic graduate classes in quantum mechanics to teach students the essentials of DFT. Practical DFT training often focuses on how to use packages, with less emphasis given to understanding the underlying methods and their limitations. The CoMET program directly addresses the need for DFT literacy and involves faculty from three Colleges and six different departments across science and engineering at Penn State. Through CoMET, we have established a graduate minor in Computational Materials, which highlights two courses in fundamental and practical aspects of DFT. Integral aspects of this program are the DFT Café, where students and faculty meet weekly to discuss key papers in the development and use of DFT, a weekly seminar series, where students present DFT-based research, an international summer school, and a capstone project. I will discuss the deployment of the CoMET program, including hurdles that we faced, our progress so far, and the outlook for implementing elements of this program into the graduate Chemical Engineering curriculum.

535918 Software Strategies for Increasing Throughput and Reproducibility While Lowering Cognitive Load in Molecular Simulation

Abstract:

Learning molecular simulations and reproducing someone else's molecular simulations are both hindered by software dependency issues: Acquiring the version of Software A to accomplish Task B results in a chain of installation and compilation steps that require significant expertise to navigate nuances associated with a user's specific system architecture, operating system, and software stack. The effort associated with setting up and resolving software conflicts is one example of `cognitive load' that burdents novice molecular simulators disproportionately. In order to make molecular simulations a more mainstream engineering tool, software developers can use the minimization of cognitive load as a guiding principle to inform design decisions.

In this work we discuss three examples in molecular simulation infrastructure development aimed at lowering cognitive load. First, we discuss the use of Docker containers for deploying software stacks to new computational scientists. Maintaining tutorial information and software dependencies through Docker files ensures that versions always match, and are foundational to campus-wide computing efforts now in their third year at Boise State University. Second, we detail continuous integration and regression testing during the development of a simulation plugin to model epoxy crosslinking dynamics. By automatically catching software changes that influence the scientific correctness, developers can focus their efforts on optimizing performance more effectively. Finally, we show examples of using signac-flow to manage the execution of studies with thousands of jobs. By automating post-processing and handling of job scripts, users can focus efforts on understanding their simulation results instead of tracking and manually editing thousands of files. In concert, we demonstrate how an emphasis on lowering cognitive load leads to shorter training times for new scientists and improved transfer of information between researchers deriving from enhanced reproducibility.

536063 Reproducible Computational Workflows with Signac

Abstract:

Researchers in computational science are regularly posed with the challenge of managing and analyzing large, heterogeneous, and highly dynamic data spaces. We present *signac*, a non-intrusive, open-source Python framework that enables researchers to efficiently operate on primarily file-based data spaces while keeping track of all relevant metadata. The *signac* framework provides all components required to create a well-defined, collectively accessible data space and to implement reproducible workflows. The software is designed to be highly modular, decoupling its data and workflow management components so as to minimize the effort required for integration into existing workflows. The serverless data management and lightweight workflow model ensure that workflows are just as easily executed on laptops as in high-performance computing environments. The signac approach not only increases research efficiency, it also improves reproducibility and lowers barriers for data sharing by transparently enabling the robust tracking, selection, and searching of data by its metadata.

538303 Making Data-Driven *in silico* Research a Mainstream Chemical Engineering Tool

Abstract:

Trial-and-error approaches are increasingly ill equipped to meeting the complex challenges involved in the discovery and design of next-generation chemistry and materials. Our work recognizes the great opportunities that are arising with the shift towards data-driven *in silico* research and a rational design paradigm. These approaches are poised to mitigate the inefficiencies, shortcomings, and limitations of traditional trial-and-error research. However, the notion to utilize modern data science in the chemistry context is so recent that much of the basic infrastructure has not yet been developed, or is still in its infancy. The existing tools and expertise tend to be in-house, specialized, or otherwise unavailable to the community at large. Data science is thus in practice beyond the scope and reach of most researchers in the field. Our work aims to chart new paths in this area by creating an open, general-purpose software ecosystem designed to overcoming this situation, filling the prevalent infrastructure gap, and thus making data-driven research a viable and widely accessible proposition. Our software ecosystem fuses *in silico* modeling (in particular computational quantum chemistry), high-throughput screening techniques, and Big Data analytics into an integrated research infrastructure. We have been developing the necessary methods, algorithms, protocols, and codes, and assembled them in three loosely connected program suites: ***ChemHTPS*** provides an automated platform for the virtual high-throughput screening of compound and material candidate libraries as well as reaction networks; ***ChemBDDB*** offers a database and data model template for the massive information volumes created by data-intensive projects; and ***ChemML*** is a machine learning and informatics toolbox for the validation, analysis, mining, and modeling of such data sets.

537683 Open Chemistry, Avogadro and Jupyter: User Friendly Frontends

Abstract:

Open, interactive interfaces employing data-centric workflows have been developed, reusing best-of-breed open source software in order to deliver an integrated platform for knowledge discovery. The Open Chemistry project serves as an umbrella, Avogadro offers a desktop molecular editor, and the Jupyter project offers a powerful electronic notebook interface, with server-side software kernels executing Python code. The JupyterLab web frontend provides a web-based interface with interactive cells where code can be edited, and panels where data can be visualized. Coupling these interfaces with a powerful data server, capable of triggering simulations, analyses, and other workflows offers a powerful capability to seamlessly execute codes from a pre-configured environment, store data, and share the results. The use of software containers improves both modularity and reproducibility.

Extension of the Python software kernels and web interface with chemistry specific capabilities results in a software environment. The development of frontends is important to foster wider adoption, the use of reproducible electronic notebooks offers an environment where simulation steps can be shared accurately, and robustly. The use of collaborative editing capabilities offers the peer-to-peer review of the complete process from initial chemical coordinates to final results. A data-centric environment, with Python scripting, web programming interfaces, and modern web visualization widgets creates an environment that can be used for teaching, research, and wider dissemination.

The development of an open ecosystem of tools will be described, with open programming interfaces that feature working reference implementations. The best way to make molecular simulation more mainstream is to embrace these modern, data-centric user interfaces, use modern, documented file formats for exchange, and to make all steps of the process transparent. Software containers are used to package computational codes, and to run them using cloud or high-performance computing resources. Improved standardization of file formats will make it easier to add more codes, and open systems enable extension by simulation code developers, researchers, developers, and instructors.

538283 COMSOFT Workbench: Tools for Efficient Coarse-Grained Modeling of Soft Materials

Abstract:

The objective of the Materials Genome Initiative (MGI) is a new paradigm enabling accelerated materials design via computation leading to a shortening of the time and cost needed to develop new materials for commercial applications. An essential element to meeting these objectives for polymers and related soft materials is the development of a coherent and progressive framework for carrying out coarse-grained modeling at mesoscopic length scales between the atomistic and continuum. Such a framework requires meeting the theoretical challenge of describing the representation and interoperability of materials data and the interoperation of modeling systems at multiple length and time scales, as well as the practical challenge of creating a computing environment that realizes its instantiation.

In this seminar, we describe recent progress on â€œCOMSOFT Workbench[1]â€ a NIST Materials Genome project to provide algorithms, tools and data to better support coarse-grained modeling of soft materials. The Workbench application is built on top of the OpenChemisty[2] project. In the Workbench project, we add several new features aimed at simplifying the task of multiscale modeling. The first is a Python scripting library, COMSOFT.py, which supports a multiscale description of materials in which atomic and coarse-grained descriptions are integrated in a single hierarchical data structure. Using this, we build scripts for automating complex tasks and then link these into the application through the GUI. The primary focus of the work is the development of algorithms for automating the parametrization of coarse-grained force-fields using progressive methods. Two algorithms will be supported in the initial release, Iterative Boltzmann Inversion (IBI) and the recently developed temperature transferable algorithm based on energy renormalization[3]. The final element of the application is that of reference data. The NIST Materials Data Curation System (MDCS)[4] is integrated into the environment to support ontology based database descriptions enabling integrated data curation and the progressive building of material reference libraries.

References

1. COMSOFT Workbench on GitHub, https://github.com/usnistgov/COMSOFT-Workbench
2. OpenChemisty on GitHub, https://github.com/OpenChemistry
3. Xia *et al.*, J. Phys. Chem. B, 2018, 122 (6), pp 2040â€"2045, DOI: 10.1021/acs.jpcb.8b00321
4. Materials Data Curation System (MDCS) on GitHub, https://github.com/usnistgov/MDCS

538523 MoSDeF: Molecular Simulation and Design Framework for Transparent, Reproducible, Usable By Others, Extensible Simulations (TRUE)

Abstract:

Reproducibility in scientific research has become a prominent issue [1]. Computational scientists, along with the rest of scientific community, are grappling with the central question: How can a computational study be performed in such a way that it can be replicated by others? To this end, we discuss the Molecular Simulation and Design Framework (MoSDeF) [2], a suite of Python-tools designed to facilitate reproducible molecular simulations, with an emphasis on the ability to perform large scale screening of structural/chemical space. The core tool, mBuild [3,4], is used to generate configurations of chemical systems of interest. mBuild leverages the concept of generative modeling, where complex systems can be constructed from smaller, interchangeable components, allowing for simple parameterization of structures. Repetitive structures like polymers, crystals, spherical or planar tiling patterns can be expressed declaratively. These features of mBuild allow users to programmatically vary parameters for a family of systems (e.g., polymer chain length) as well as vary interchangeable components (polymer subunits, crystal basis particles, etc) with minimal changes to the underlying code. mBuild integrates with Foyer [5,6] a tool for defining force field parameter usage and applying force fields to molecular systems (i.e., atom-typing). Foyer provides a force field and simulation engine agnostic method for defining parameter usage that relies upon SMARTS [6] based annotations of chemical context and overrides statements to set rule precedence. As such, these annotations provide both human and machine readable documentation of parameter usage, reducing ambiguities as to how parameters should be used and removing rule order as a source of error. By separating rules from the code to evaluate them, force field files can be constructed, modified, version controlled, and easily disseminated. We have coupled this framework with the workflow management tool Signac-flow8 enabling MoSDeF to perform large scale parameter screening, capturing the exact MoSDeF procedures, inputs, and other relevant metadata. Combined, these tools provide a framework for performing simulations that are Transparent, Reproducible, Usable by others, and Extensible (TRUE). We demonstrate the capabilities of MoSDeF with a case study examining the tribological properties of functionalized monolayer films, screening over chain length and terminal group functionality, to discover structure/property relationships that should aid in the design of better-performing films.

References

[1] Baker, M. 1,500 Scientists Lift the Lid on Reproducibility. *Nature* **2016**, *533*(7604), 452–454.

[2] "MoSDeF" [Online]. Available: https://github.com/mosdef-hub.

[3] C. Klein, J. Sallai, T. J. Jones, C. R. Iacovella, C. McCabe, and P. T. Cummings, "A Hierarchical, Component Based Approach to Screening Properties of Soft Matter", Foundations of Molecular Modeling and Simulation, **2016**, pp. 79-92.

[4] "mBuild" [Online]. Available: https://github.com/mosdef-hub/mbuild.

[5] Iacovella, C. R.; Sallai, J.; Klein, C.; Ma, T. "In Idea Paper: Development of a Software Framework for Formalizing Forcefield Atom-Typing for Molecular Simulation", 4th Workshop on Sustainable Software for Science: Practice and Experiences (WSSSPE4), **2016**.

[6] "foyer" [Online]. Available: https://github.com/mosdef-hub/foyer.

[7] "SMARTS" [Online]. Available: http://www.daylight.com/dayhtml/doc/theory/theory.smarts.html

[8] "Signac-flow" [Online]. Available: https://bitbucket.org/glotzer/signac-flow.

529746 Dissipative Particle Dynamics Simulations of Anion Exchange Membranes

Abstract:

Anion exchange membrane (AEM) is composed of a polymer backbone tethered with ion exchange side chains, which transports ions and water within its nano-segregated hydrophilic domains while being hydrated. Studies of AEM in the past few decades show that AEM is a competitive alternative to conventional proton exchange membranes (PEM) in fuel cell (FC) application, mainly because of the alkaline environment allowing the use of non-noble metal for electro-catalysts.

Here, dissipative particle dynamics (DPD) simulation is used to explore the relation between the diffusivity of mobile components and mesoscopic structures of AEM in terms of different hydration levels, ion exchange capacities, and types of copolymers and functional groups. The provided fundamental information is useful for designing more versatile AEMFCs.

Previously devised strategies for studying PEM are transferred into this AEM work. Scale-bridging parameterization[1] determines the pairwise potential between coarse grained (CG) beads by mapping to bulk properties of the bead components. Neighbor and second-neighbor bonds[1] are used to regulate CG molecules in order to preserve their configurations on the atomistic level. Short-range Morse potential is used to mimic the transport mechanisms of anions and the association-dissociation equilibrium[2] between anions and the functional groups, along with their explicit electrostatic interactions. This methodology shows promising results on modeling proton conductivity in sPS[3] and Nafion[4], and its performance on modeling AEM will be discussed.

References

[1] **M. T. Lee**, R. F. Mao, A. Vishnyakov, and A. V. Neimark, J. Phys. Chem. B **120** (2016) 4980-4991.

[2] **M. T. Lee**, A. Vishnyakov, and A. V. Neimark, J. Chem. Theory Comput. **11** (2015) 4395-4403.

[3] **M. T. Lee**, A. Vishnyakov, and A. V. Neimark, J. Chem. Phys. **144**, 014902 (2016)

[4] A. Vishnyakov, R. Mao, **M. T. Lee**, and A. V. Neimark, J. Chem. Phys. **148**, 024108 (2018)

528873 Computational Investigation of Water Desalination across Nanofiltration Membranes Using Advanced Sampling Techniques

Abstract:

Freshwater availability and sustainability is a major challenge facing humanity in upcoming decades. A potent approach to address this challenge is seawater desalination, which, despite its several outstanding issues, has the potential to provide a sustainable framework for freshwater production. Currently, membrane processes such as reverse osmosis (RO) and nano filtration (NF) are the most widely used desalination approaches both in terms of installed capacity and annual growth. Membrane separation normally requires applying driving forces such as pressure or electric field in order to overcome the natural osmotic pressure gradient between pure and salty water. There is, however, significant room for improvement of the existing technologies both in terms of energy efficiency and economic affordability, making studies of filtration an active area of research.

In this project, we use molecular dynamics (MD) simulations and advanced sampling techniques to investigate ion transport across nano-porous graphene membranes. The two most important metrics for efficiency of a water desalination process are the water flow rate and the salt rejection. We are particularly interested in using MD and a path sampling technique known as forward-flux sampling (FFS) to predict the flux of water molecules and various anions such as Chloride (Cl^-) and Nitrate (NO_3^-) and to inspect their dependence on process variables such as temperature and the applied pressure gradient. We will also investigate the effect of functionalization and hydrophobicity of nanopores on the flow rate and salt rejection. Using an advanced sampling technique such as FFS enables us to estimate fluxes that are too small to quantify in a conventional MD simulation, and can therefore be utilized for rational design of more efficient filtration processes.

537687 Combining Biased Sampling and Markov State Models to Characterise the Assembly and Exchange Dynamics of Molecular Materials in Solution

Abstract:

Achieving an efficient and scalable production of advanced molecular materials is a central issue in the development of novel chemical processes. Leveraging self-assembly to efficiently produce molecular materials such as crystals, porous matrices or non-covalent polymers has the potential to impact a variety of applications, including pharmaceutical manufacturing, separations, sensing and catalysis. In this context, understanding the molecular-level fundamentals of self-assembly, growth, and dissolution processes is key.

A representative example is provided by molecular crystals. It is known that both the particle morphology and bulk structure of molecular crystals depend on molecular-level kinetics, which in turn are impacted by the composition of the parent phase and by the conformational complexity of their constitutive units [1]. In this case obtaining insight into molecular-level thermodynamics, mechanisms, and kinetics of crystal nucleation and growth, while explicitly accounting for the composition of the liquid phase and the structural complexity of the building units is essential to understand, model and design efficient crystallization processes.

Molecular dynamics (MD) simulations naturally provide the spatial resolution necessary to achieve this level of detail. However in order to apply MD to investigate assembly processes we need to bridge the gap between computationally accessible timescales - of the order of ~Î¼s-ms - and the characteristic timescale of relevant events â€“ often exceeding the seconds. A viable approach at tackling this challenge is the application of biased sampling methods such as Well-Tempered Metadynamics (WTMetaD) [2] to achieve an extensive exploration of the configuration space and allow for an efficient calculation of transition rates.

In this contribution we provide an overview of the application of WTMetaD and Markov State Models (MSM) [3] to the investigation of self-assembled molecular materials in solution. We discuss three classes of systems of increasing complexity.

We begin by analysing the conformational isomerism of ibuprofen in solution and at the crystal/solution interface. In this case we employ WTMetaD to obtain free energy surfaces associated to ibuprofen conformational transitions. Free energy basins allow to identify a set of states used build a MSM. Transition rates between states are computed from extensive unbiased sampling or by reconstructing the transition time distribution from state to state infrequent WTMetaD simulations [4, 5]. This analysis uncovers how the interplay between solute, solvent and interface structure leads to significant variations in the conformers equilibrium probability, in the relaxation rates of the conformer population, and in the dominant mechanism of conformational rearrangement of ibuprofen.

We continue by discussing the construction of a thermodynamic model for describing the conformational isomerism of structural building units (SBUs) involved in the early stages on the nucleation of MIL101(Cr) metal-organic framework [6]. In this case, by running extensive WTMetaD simulations of the formation of SBUs from smaller precursors we unveil a large number of potential structural isomers of SBUs. We analyse the ensemble of sampled structures by identifying a subset of these SBUs that retain a configuration compatible with that adopted in the MIL101(Cr) bulk. By systematically building equilibrium models for SBUs formation at increasing ionic strength we show that ions tend to screen ligand-metal interactions and, above a certain threshold, ions favour the formation of crystal-like SBUs by establishing salt bridges.

Finally, we discuss the application of infrequent WTMetaD to compute rates and identify molecular mechanisms associated with monomer exchange in structural variants of 1,3,5-benzenetricarboxamides (BTA), a family of molecules capable of assembling into water-soluble supramolecular polymers [7]. In this case molecular simulations capture supramolecular dynamics consistent with experimental observations, providing the information necessary to link structural and dynamical features of BTA variants. These findings lead to the construction of a general kinetic model capable of reproducing key features of experimental FRET mixing essays for non-covalent BTA-like fibres.

To conclude, in this contribution we highlight how the systematic application of enhanced sampling methods to discover relevant intermediate states, compute transition rates, and build kinetic models, opens up the possibility of systematically broadening our understanding of mechanisms, kinetics, and thermodynamics of a variety of molecular assemblies in the liquid phase ranging from molecular crystals to supramolecular fibres.

References

1. M. A. Lovette, A. B. Browning, D.W. Griffin, J. P. Sizemore, R. C. Snyder, M. F. Doherty, *Ind. Eng. Chem. Res*, 2008, *47*, 9812
2. A. Barducci, G. Bussi, M. Parrinello, Phys, Rev. Lett., 2008, 100, 020603
3. G. R. Bowman, V. S. Pande, Springer, USA
4. P. Tiwary and M. Parrinello, Phys. Rev. Lett., 2013, 111, 230602
5. M. Salvalaglio, P. Tiwary, M. Parrinello, J Chem Theory and Comput, 2014, 10 (4), 1420-1425
6. D. C. Cantu, B. P. McGrail, and V. A. Glezakou, Chem. Mater. 26, 6401, 2014
7. D. Bochicchio, M. Salvalaglio, G.M. Pavan, 2017. Nat. Commun., *8*.

523402 In silico Prediction of Structural Properties of Racemic Porous Organic Cage Crystals

Abstract:

In silico Prediction of Structural Properties of Racemic Porous Organic Cage Crystals

Yang Liu and David S. Sholl

School of Chemical and Biomolecular Engineering, Georgia Institute of Technology, USA

Corresponding e-mail: yliu738@gatech.edu

Abstract

Porous organic cage (POC) solids have shown promising properties and attracted significant attention since their discovery. Chirality within POC materials plays an important role in controlling the crystalline assembly of these porous organic molecules. Most previous studies focus on racemic POC crystals that are mixtures of enantiopure cages. However, POC crystals can also be synthesized from racemic mixtures of diamines. A low-cost synthesis route for POC crystals with racemic diamines as reactants has been adopted in our previous study,[1] and recently investigated again by Cooper and coworkers[2]. In the previous work, we reported a new CC3-based crystal synthesized from a racemic mixture of (1R,2R)- and (1S,2S)- DACH and TFB, and found this new crystal has more improved properties relative to crystals synthesized from enantiomerically pure DACH. It is challenging to solve this new racemic crystal structure experimentally because only subtle differences exist between the new racemic structure and previously reported CC3 structures.

Here we introduce and develop an *in silico* prediction method that combines electronic structure calculations and molecular-level calculations to predict the atomic structure for racemic POC crystals. Our procedure starts from enumerating types and corresponding concentrations of cage molecules. Then, the packing of those individual porous organic cage molecules within the racemic crystal are studied and understood. Furthermore, lattice model representations of the racemic crystal are constructed and predicted with (MC) simulation. Finally, by expanding the lattice model into atom-level detail, atomically detailed CC3-racemic crystal models can be created. Agreement between experimentally measured and simulated XRD spectra of CC3-racemic crystal supports the validity of our theoretical prediction method.

Reference

1. Zhu, G.; Hoffman, C. D.; Liu, Y.; Bhattacharyya, S.; Tumuluri, U.; Jue, M. L.; Wu, Z.; Sholl, D. S.; Nair, S.; Jones, C. W.; Lively, R. P., Engineering Porous Organic Cage Crystals with Increased Acid Gas Resistance. *Chemistry* **2016,** *22* (31), 10743-7.

2. Slater, A. G.; Little, M. A.; Briggs, M. E.; Jelfs, K. E.; Cooper, A. I., A solution-processable dissymmetric porous organic cage. *Molecular Systems Design & Engineering* **2018,** *3* (1), 223-227.

537935 NMR Relaxation from Molecular Simulations: Study on Bulk Hydrocarbons and Water

Abstract:

Nuclear magnetic resonance(NMR) relaxometry is used to get properties like pore size distribution, fluid distribution, and wettability in reservoir exploration. However, the interpretation of the NMR log data relies on classical models, that make important assumptions on the nature of the fluid and environment. Here we use molecular dynamics and NMR theories to seek a fresh perspective on interpreting relaxation mechanisms. In this talk, I will discuss the interpretations of dipole-dipole and spin rotation relaxation mechanisms in bulk systems. The simulated relaxation and diffusivity are in excellent agreement with experiments. However the results are consistent with the classical models of relaxation that assumes particles are spherical and interactions are isotropic. I will further discuss the effects of internal motions and molecular geometry on relaxation. Interestingly the relaxation behavior of neopentane closely conforms to classical behavior. Time permitting, I will also discuss our ongoing work examining NMR relaxation in alkane-polymer mixtures.

Densities and Viscosities of H_2S at Elevated Pressures and Temperatures Using Molecular Dynamics Simulations

Dr. Abhinav Verma, Shell India Markets Pvt. Ltd, Bangalore, Karnataka, India
Dr. Rajdeep S. Payal, JNCASR, Bangalore, Karnataka, India
Dr. S. Balasubramanian, JNCASR, Bangalore, Karnataka, India
Dr. Indranil Rudra, Shell India Markets Pvt. Ltd, Bangalore, Karnataka, India

Introduction

Hydrogen Sulfide (H_2S) is a hazardous chemical compound, known to cause a wide variety of adverse health effects. It is naturally found in crude oil, produced water, natural gas and hot springs. It can also be formed by anaerobic decomposition of biomass by bacteria. It is flammable, toxic and corrosive in nature. Since it occurs in crude oil and natural gas, its removal from the petroleum compounds poses a significant challenge to the petroleum industry. The amount of H_2S varies substantially between fluids from different fields and reservoirs and it becomes even more critical when the concentration of H_2S is high. There are a variety of methods to remove the H_2S from the crude oil or natural gas e.g. Claus process. But these processes are expensive and not environmentally friendly. In order to address these issues, currently H_2S is reinjected back into the reservoir and disposed into an underground formation. The process of reinjection happens at elevated pressures and hence the physical properties such as density and viscosity become important. Current state of the art method like TRAPP predicts viscosity of H_2S at high pressure with deviation from experimental values up to 42% at 750 bar and 323.15 K (experimental value from Alberta Sulphur Research Ltd (ASRL) dataset). It has also been observed that TRAPP non-systematically under predicts or overpredicts viscosity of H_2S depending on the pressure and temperature conditions. For example, at 373.15 K the predicted viscosity deviates between -23% (103 bar) to +24% (1000 bar) relative errors. An internal study suggests that no method in Unisim design can be recommended for high pressure viscosity prediction of H_2S/CO_2 gases. The conditions for which experimental data for H_2S is available are shown in Figure *1*.

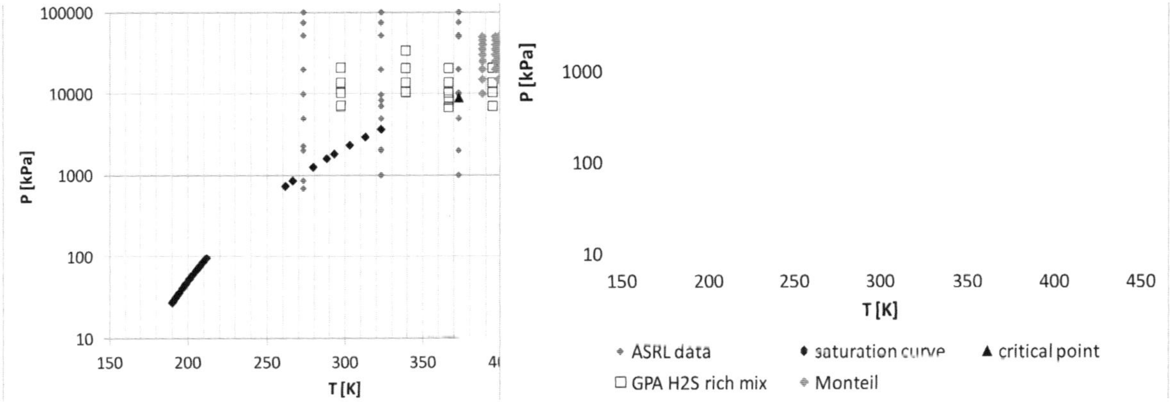

Figure 1: Available experimental data for pure H2S and H2S rich mixture viscosity, from different sources [1-3]

Molecular modelling and simulations can play a vital role in such scenarios as it is physics based and in principle can be predictive if a reliable intermolecular potential (also popularly known as force-field) is available. However, in practice such a force field is rare which is reliable over a wide range of pressure, temperature conditions. Therefore, developing new force-fields and improving existing force-fields are important from method development standpoint. For industrial application detailed studies of reliability of existing force-fields for a given molecular system are important. Molecular and atomic level insight provided by the simulations can be instrumental in designing the new compounds and methodologies for desulfurization of crude oil and natural gas, including a proper prediction of fluid properties.

Methods

Overview of Previous Molecular Dynamics Studies

Molecular dynamics (MD) simulations is an apt choice to understand the behavior of a compound (or a mixture of compounds) under variety of thermo-physical conditions. Results of an MD simulations are governed by the choice of the force field employed. First MD simulations of H_2S were done the group of Jorgensen in 1986 [4]. They developed the three-site point charge model for H_2S. This force field was parameterized against the liquid densities and enthalpy of vaporization. To achieve this, they increased the partial atomic charges on each site resulting in the large dipole moment of H_2S in the gas phase (2.1 D) as compared to the experiment (0.97 D). This resulted in the poor prediction of the other properties such as vapor pressure. Later in 1989 Klein and coworkers developed a four-site model for H_2S[5]. In this model they have added an extra charge site at the bisector of the H-S-H angle. This model was parametrized against the monomer and dimer properties obtained using quantum chemical calculations and bulk liquid properties. This model was able to predict the static and dynamic properties of liquids quite well. This model was able to describe the orientationally disordered high temperature solid phase too. But one of the major shortcomings of this model was that it under predicts the vapor pressure. In their paper the authors have made an important note regarding the behavior of H_2S, pointing to its high polarizability. Suggesting that induction forces (environment) may play a more significant role in determining the behavior of H2S. This model was further refined by the Kristof and Liszi in 1997[6]. This was done by the fitting the experimental equilibrium vapor-liquid phase data. But this model was not able to successfully predict the behavior of a binary mixture of H_2S with CO_2.

Since it was pointed that H_2S has high polarizability, Delhommelle *et. al* [7] developed a polarizable model for H_2S. Their potential model includes electrostatic and polarizable sites in addition to Lennard-Jones sites. They stressed that such models are essential for describing the phase equilibria correctly. Riahi *et. al[8]* and Orabi *et. al [9]* also developed the Drude polarizable model for H_2S. Their potential model differs from Delhommelle *et. al* in the way polarizability is introduced via charged (Drude) particle attached to the S atom through a harmonic potential. But Nath [10] developed a simple point charge model for H2S that was able to describe the vapor-liquid phase equilibria correctly. But this model slightly under predicted the boiling point of these systems (predicted boiling point 205.8 K vs experimental value of 212.8K). Such issues were addressed by Kamath et al. [11]. They developed the four different models for the H_2S with different Lennard-Jones+ partial charge parameters. But even their models were unable to predict interaction between n-pentane and H_2S correctly leading to systematic overprediction of liquid phase mole fraction of H_2S. Recently Siepmann's [12] group has developed the Transferable Potentials for Phase Equilibria (TraPPE) (force field) for H_2S. Among the four different potential models developed by their group, the 4-site model was able to predict the thermo-physical quantities best. This model consists of four sites: both sulfur and hydrogen atoms interact via

LJ potentials, and partial charges are placed on the hydrogen atoms and an X site located along the H−S−H bisector.

The development of the accurate model for describing the behavior and fluid properties H_2S is still an ongoing process. One of the most difficult task in the simulations is to have correct prediction of the collective quantities such as viscosity and conductivity. Prediction of these quantities and behavior of H_2S under different thermo-physical conditions is of acute importance for the petrochemical industry. For this, availability of experimental data for model validation is a critical enabler. However, it has been noted in earlier work (internal) that data for pure hydrogen sulphide at high pressure are scarce and not always mutually consistent. The available experimental data was compiled from four different sources. This work has been devoted to have the correct estimate of the viscosity of the H_2S under various temperature and pressure conditions using the MD simulations. We have used the available experimental data from different sources [1-3] for validation.

Computational Details

The force field developed by Nath [10] is simple 3-site model. In a 3-site model each atom gets a point charge and the associated Lennard-Jones parameters. It is possible to have higher-site model in with the extra sites are artificially introduced to account for electrostatic distribution around the molecule. Due to the fixed nature of the required sites, the 3-site models are easier to control and simulate. Also, with respect to computational cost, the number of required interactions is lower than higher-site models and therefore the simulations can be performed at a relatively lesser computational cost. The 3-site model developed by Nath [10] can reproduce the experimental densities and structure quite well. For H_2S, the predicted dipole moment was about 13% higher than the experimental value. The dynamical properties such as viscosity obtained using this model are erroneous. Due to the simplicity of implementation of 3-point charge model, we used this model as our framework and readjusted the LJ parameters against quantum chemical calculations. LJ parameters were calculated using the MP2 level of theory using the G09 software package[13]. This set of new parameters (shown in Figure *2*) was used for the further MD simulations.

All atom MD simulations were performed in both NPT and NVT ensembles for the H_2S system at various temperature and pressure conditions which were chosen depending on availability of experimental data. Long-range electrostatic interactions were computed using the particle−particle particle−mesh (PPPM) method. The real space cutoff for pairwise interaction in real space was 10 Å with the initial box size of ~28 Å. Long-range corrections for the energy and pressure were also applied.

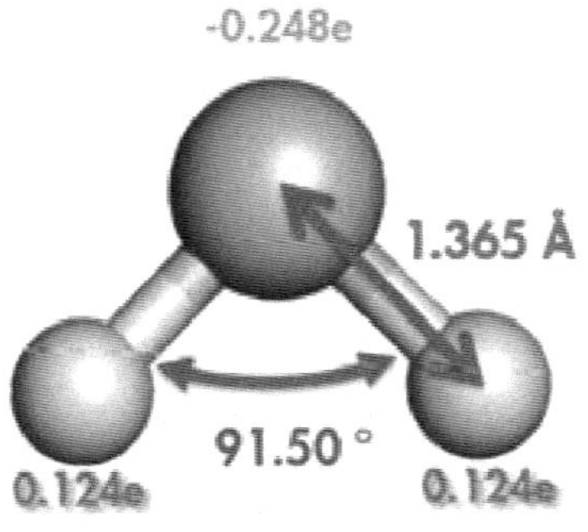

Element	ε (kcal/mol)	σ (Å)	Mass (g/mol)
S	0.50585	3.72	32.066
H	0.00785	0.98	1.00794

Figure 2: H_2S and related forcefield parameters

Equations of motion were integrated using the velocity-verlet algorithm, with 1 fs integration time step. The S–H bonds were constrained using the SHAKE algorithm. Cross interactions between atoms were calculated using the standard Lorentz–Berthelot rules. The temperature and pressure of the systems were maintained by using a Nose–Hoover thermostat and barostat respectively, with a damping factor of 500 fs and 1 ps respectively with cubic periodic boundary conditions.

Each system consisted of 256 H_2S molecules. Initial configurations were generated using Packmol and minimized for energy first followed by a short NVT equilibration and then by a long NPT run of 10 ns to get the equilibrium volume of the system. This equilibrated system was used to generate the analysis trajectory in the NVT ensemble. Shear viscosity was calculated through the equilibrium Green–Kubo relation using the full stress (pressure) tensor,

$$\eta(t) = \frac{V}{10k_bT}\int_0^t \langle Tr[P(t)P(t+t')]\rangle\, dt \qquad \text{Eq. (1)}$$

where V and T are simulation box volume and temperature respectively. P is the symmetric, traceless part of the pressure tensor. Angular brackets represent a time average. Using equation (1) yields better results than one with only off diagonal components[14]. The viscosity of the system was calculated in same fashion as described in our previous work[15].

All the simulations were done using the software package LAMMPS. VMD [16] was used to visualize the simulation trajectory and create the simulations snapshots. All the analysis was done by the homegrown FORTRAN 90 codes reported earlier[17].

Results

Densities of H2S from MD simulations

Calculated densities for H_2S at various state points are tabulated in Table *1* and are in good agreement with the experimental densities. Our simulations slightly over predict the density as compared to the experiments. It may be noted that the data is coming from various sources and therefore might itself be subject to different errors. This model fails to estimate the density at critical point of H_2S (T = 373K P = 90 Bar) as it overpredicts by 13% (expt 0.477 g/cc) (data point not shown in the Table *1*)

Table 1: Experimental and calculated densities of H2S

Temp (K)	Pressure (kPa)	$Density_{exp}$ (g/cc)[20]	$Density_{sim}$ (g/cc)	Density (%)
217	124	0.943	0.952	0.95
333	4234	0.680	0.691	1.61
373	40327	0.715	0.727	1.80
373	72447	0.784	0.789	0.65
413	13172	0.272	0.274	0.73

We performed ab-initio MD NVT simulations using CP2K starting with experimental density at 217 K, 1.2 atm and at 333 K, 40 atm. Figure *3* shows the comparison of the radial distribution functions obtained from the classical MD simulations (using the force field parameters shown in Figure *2*) and ab-initio MD simulations. The radial distribution function is related to the structure of H_2S and the two calculations are in good agreement with each other. This suggests validity of the new set of parameters for capturing the overall structural effects at two different pressures and temperatures.

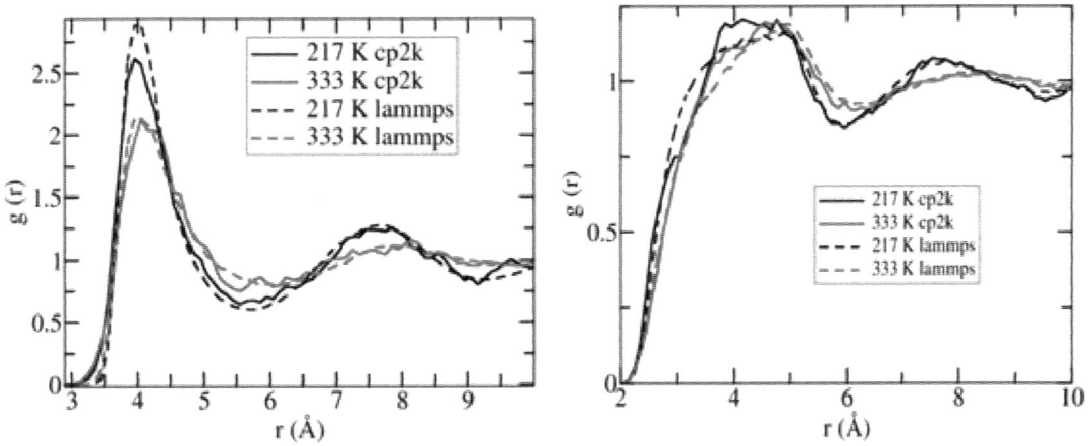

Figure 3: Radial distribution function between (a) S-S and (b) H-S in the H_2S molecule.

We also compare the structure of clusters of few H_2S (dimer, trimer, tetramer and pentamer) obtained via MD simulations with quantum chemical calculations. Our MD simulations were able to predict the clusters obtained from the quantum chemical calculations. We also calculated change in net dipole moment as function of cluster size. The dipole moment of H_2S monomer in gas phase is 0.97 Debye. In literature [18]higher values of dipole moment are used in non-polarizable models. It is argued that this allows indirectly to account for polarization effect. For H_2S dimer dipole moment of 1.62 D and for trimer dipole moment value of 1.75 D was used[19]. The dipole moment values for H_2S clusters lie in the range expected for non-polarizable potential models reported in literature (between 1.4 D to 2.1 D)[18].

Density and Viscosity calculations for H_2S

A combination of high pressure data for both viscosity and density is scarcely available in the literature in case of H_2S (Figure 1). We used Dortmund databank [20] to extract experimental data that were available. We used our earlier reported protocol [15] to predict densities and viscosities at these selected conditions.

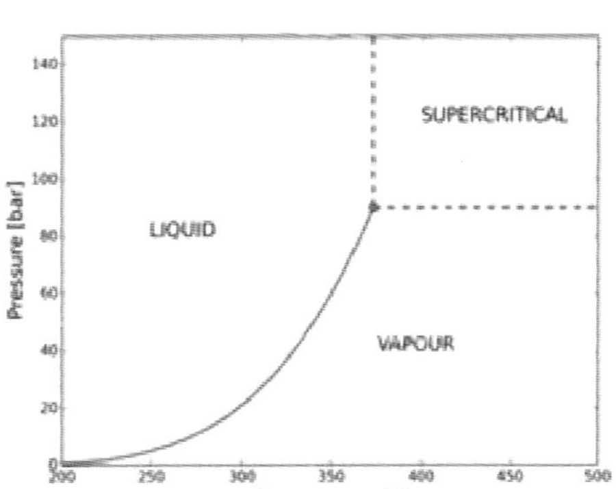

Figure 4: Illustrative phase diagram of pure H_2S. Solid phase is not shown.

In previous section we used experimental density of liquid phase and vapor phase (at T≥Tc, P>Pc) H_2S to validate our potential model. Illustrative phase diagram of pure H_2S is shown in Figure *4*. The critical point is observed at 373K and 90 Bar. The predicted and experimental densities for 388K and 413K are also shown in Figure *5* that lie in the supercritical phase of H_2S.

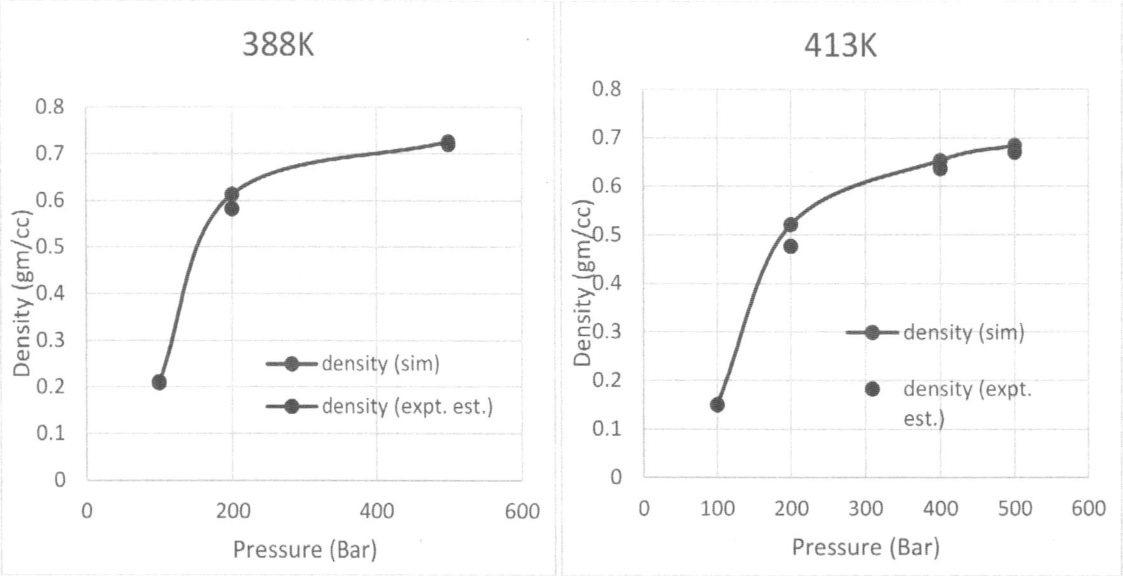

Figure 5: Density of H_2S at 388K & 413K and pressures ranging from 100-500 Bar

Table 2: Density of H_2S in liquid phase

T [K]	P [bar]	Density (sim) [g/cc]	Density (expt.) [g/cc] Error! Reference source not found.0]	error %
273	29.76	0.841585	0.83667	-1%
273	39.77	0.840466	0.8385	0%
273	49.81	0.835197	0.84028	1%
273	59.85	0.843556	0.84202	0%
273	69.86	0.851394	0.84375	-1%
273	79.84	0.833907	0.84545	1%
273	89.88	0.840667	0.84711	1%
273	99.9	0.864725	0.84876	-2%
273	150.07	0.855523	0.85627	0%
273	200.01	0.867615	0.8637	0%
273	249.97	0.868903	0.8707	0%
273	299.91	0.878353	0.8773	0%
273	349.98	0.880826	0.88357	0%
273	400	0.893331	*0.88953*	0%

The force-field can capture the densities of the system at a wide range of pressures and temperatures. It should be noted that state points are in the supercritical state (Tc = 373K Pc = 90 Bar) and are predicted within ~10% relative errors. It is encouraging that the forcefield can capture the densities in the supercritical phase space. We have also calculated densities of H_2S in the liquid phase at different pressures and 273 K (Table *2*). Density at liquid phase is well reproduced by our model. The current potential model (force field) doesn't perform well for vapor phase with low density values (density from simulation have >6% deviation from experimental value). We believe it is due to high polarizability of H_2S in gas phase (3.63 $Å^3$) which is two times more than H_2O. It is difficult to capture this effect for the simple non-polarizable models.

We further calculated the viscosity using the equilibrated boxes form molecular dynamics simulations and compared them against the experimental data. The results are reported in Table *3*.

Table 3: Viscosities of H_2S at elevated temperature and pressures (supercritical phase)

T [K]	P [Bar]	viscosity (sim) [mPas sec]	viscosity expt [mPas sec]	error %
388	100	0.0215	0.0777	72%
388	200	0.0800	0.1160	31%
388	500	0.1500	0.1791	16%
398	200	0.0690	0.0957	28%
398	400	0.0910	0.1401	35%
398	500	0.1017	0.1553	35%
413	100	0.0220	0.0380	42%
413	200	0.0671	0.0857	22%
413	400	0.0959	0.1192	20%
413	500	0.1055	0.1340	21%

It is encouraging to see that the predicted viscosities follow the same trend when compared to experiment. All predicted viscosities are underpredicted and the force-field can be further evaluated to understand the underlying reasons. However, all predicted viscosities are within ~40% relative error to experiment. The conditions for these vary a significant range of temperature (388K-413K) and pressures (100-500 bar). At these conditions, our results are better compared to Unisim design (Peng-Robinson Model; data not shown) and can be used reliably in engineering applications during the reinjection of H_2S. Less than satisfactory performance of Unisim Design at high temperature and pressure in reproducing experimental viscosity of acid gases was observed. The input of molecular simulation is minimal, and essentially reduces to the knowledge of an accurate molecular model. This methodology can be a valuable tool to predict gas viscosities at elevated temperatures and pressures.

Summary

We have modified the LJ parameters of the force field developed by Nath et al. [10] and have used it for studying the physical properties of H_2S at various elevated temperature and pressure conditions. Such combinations are representative of reinjection process used to deal with excess H2S recovered from gas fields that have high concentrations of H_2S. The force-field is a simple 3-point model for a highly polarizable molecule.

In our study, we find that the densities of the H_2S at temperatures ranging from 388-413K and pressures ranging from 100-500 bar are reasonably predicted (within 10% relative errors to experiment). Such a deviation is in reasonable agreement with the performance of generally applicable engineering thermodynamic models used in for example flowsheet simulation tools like UniSim Design. All predicted densities are lower than the experimentally measured densities. To improve the prediction, we may need to resort to explicit polarizable force field models albeit at a significant increase in computational cost. Some of these uncertainties associated with experimental data is discussed in detail in ref [19]. Further we find that viscosities are also underpredicted by about ~35% relative error. For very low viscosities, the errors are higher (over 50% relative errors). This could partly be attributed to the accuracy of experimental data, variation of source and measurement techniques. Thus, we find that a very simple (modified) 3-point model could be used to predict densities and viscosities of H_2S in the higher density regime. Three site models are advantageous as the simulations are less compute intensive than the higher site models and are easier to control. Further mixing of force-fields is easier when different molecules are added into the simulation mixture, e.g., CO_2, CH_4, etc. The errors in viscosity is higher than desirable in case of low density state points. An explicit polarizable model can increase the accuracy of the predictions but will be restrictive in studying mixing of different molecules (e.g. H_2S and CO_2 or CH_4). Further studies on polarizable models can confirm the usability and accuracy of predictions. Explicit consideration of polarization is needed for improved accuracy in the vapor phase at low pressure (where the H_2S has low density).

References

1. Runovskaya, I.V., A.D. Zornin, and G.G. Devyatykh, *Viscosity of Liquefied Volatile Inorganic Hydrides of Elements of Mendeleev Periodic Groups III-VI.* Russ. J. Inorg. Chem, 1970. **15**: p. 1338-1339.
2. Steele, B.D., D. McIntosh, and H. Archibald E, *Die Halogenwasserstoffsäuren als leitende Lösungsmittel*, in *Zeitschrift für Physikalische Chemie.* 1906. p. 129.
3. Hennel, J., K. Krynicki, and *Viscosity of Liquid Hydrogen Sulfide.* Acta Phys.Pol., 1959. **19**: p. 523-526.
4. Jorgensen, W.L., *Intermolecular potential functions and Monte Carlo simulations for liquid sulfur compounds.* The Journal of Physical Chemistry, 1986. **90**(23): p. 6379-6388.
5. Forester, T.R., I.R. McDonald, and M.L. Klein, *Intermolecular potentials and the properties of liquid and solid hydrogen sulphide.* Chemical Physics, 1989. **129**(2): p. 225-234.
6. Kristóf, T. and J. Liszi, *Effective Intermolecular Potential for Fluid Hydrogen Sulfide.* The Journal of Physical Chemistry B, 1997. **101**(28): p. 5480-5483.
7. Delhommelle, J., P. MilliÉ, and A.H. Fuchs, *On the role of the definition of potential models in Gibbs ensemble phase equilibria simulations of the H2S-pentane mixture.* Molecular Physics, 2000. **98**(22): p. 1895-1905.
8. Riahi, S. and C.N. Rowley, *A Drude Polarizable Model for Liquid Hydrogen Sulfide.* The Journal of Physical Chemistry B, 2013. **117**(17): p. 5222-5229.
9. Orabi, E.A. and G. Lamoureux, *Simulation of Liquid and Supercritical Hydrogen Sulfide and of Alkali Ions in the Pure and Aqueous Liquid.* Journal of Chemical Theory and Computation, 2014. **10**(8): p. 3221-3235.
10. Nath, S.K., *Molecular Simulation of Vapor−Liquid Phase Equilibria of Hydrogen Sulfide and Its Mixtures with Alkanes.* The Journal of Physical Chemistry B, 2003. **107**(35): p. 9498-9504.
11. Kamath, G., N. Lubna, and J.J. Potoff, *Effect of partial charge parametrization on the fluid*

phase behavior of hydrogen sulfide. The Journal of Chemical Physics, 2005. **123**(12): p. 124505.
12. Shah, M.S., J.I. Siepmann, and M. Tsapatsis, *Transferable potentials for phase equilibria. Improved united-atom description of ethane and ethylene.* AIChE Journal, 2017. **63**(11): p. 5098-5110.
13. *Gaussian 09, Revision E.01.* 2009.
14. Plimpton, S., *Fast Parallel Algorithms for Short-Range Molecular Dynamics.* Journal of Computational Physics, 1995. **117**(1): p. 1-19.
15. Payal, R.S., et al., *Shear viscosity of linear alkanes through molecular simulations: quantitative tests for n-decane and n-hexadecane.* Molecular Simulation, 2012. **38**(14-15): p. 1234-1241.
16. Humphrey, W., A. Dalke, and K. Schulten, *VMD: Visual molecular dynamics.* Journal of Molecular Graphics, 1996. **14**(1): p. 33-38.
17. Verma, A., F.M. Thakkar, and I. Rudra, *Improving computational efficiency of molecular simulation-based viscosity models for diesel fuel surrogates*, in *Shell Internal Report.* 2015.
18. Ihmels, E.C., *Experimentelle Bestimmung, Korrelation und Vorhersage von Dichten und Dampfdrücken.* 2001, Universitat Oldenburg. p. 198.
19. Ihmels, E.C. and J. Gmehling, *Densities of Toluene, Carbon Dioxide, Carbonyl Sulfide, and Hydrogen Sulfide over a Wide Temperature and Pressure Range in the Sub- and Supercritical State.* Industrial & Engineering Chemistry Research, 2001. **40**(20): p. 4470-4477.
20. *Dortmund Data Bank.* 2015.

525977 Greenhouse Gas Capture: A Recent Theoretical Advancement

Abstract:

Asymmetric solutions are the rule rather than the exception in vapor-liquid (VLE) equilibria situation. This asymmetry stems from the difference in the interaction between the solute and solvent particles in the systems. The solubility of Greenhouse gases can be modelled using Henryâ€™s Law at the subcritical temperature where the temperature (T) is far below the supercritical temperature (T_c). The analytical extrapolation vapor pressure in the near supercritical region usually leads to the underestimation of pressure. In this talk, we will show just by invoking the new concept of asymmetric interaction into the systems, one is able to improve the prediction of solubility of CO_2 in room temperature ionic liquids (RTILs). Also, one can use the model to predict the solubility over a wide range of temperatures and pressures from the near subcritical to the near supercritical region with an average absolute relative deviation (AARD) of 14%. Also, the new asymmetric model is found to improve the CO_2 solubility in RTILs with an AARD of 11.44% at ambient condition with respect to the other solubility models of Greenhouse gas capture in this field.

525234 Mechanisms of Synthetic Chloride Ion Transporters in Lipid Bilayer Membranes

Abstract:

Natural lipid bilayer membranes are impermeable to many biologically relevant molecules and atomic species. Special transmembrane protein complexes embedded in lipid bilayers facilitate the transport of physiologically necessary molecules across the barrier. For this reason, when diffusion rates and concentration gradients across a cell membrane must be precisely regulated, protein complexes are often involved. Intracellular concentration of ions are commonly regulated by these complexes. Chloride channels comprise a superfamily of protein complexes specific for the chloride ion gated by one of four main mechanisms â€" membrane potential, Ca^{+2} ion gradient, ligand, or pH. Dysfunction in these channels is responsible for several serious human pathologies including cystic fibrosis. SCMTRâ€™s (Synthetic Chloride ion Membrane TRansporters) are a class of designer molecules that mimic the function of natural chloride channels, exhibiting voltage gated transport of chloride ions across artificial lipid membranes. Previously, the mechanisms by which these synthetic transporters operated were poorly understood. New molecular dynamics simulations of SCMTRâ€™s in lipid membranes of pure DOPC and 50/50 mixtures of DOPC/DOPE have uncovered the mechanisms of channel formation and voltage gating. A recently developed method, known as computational electrophysiology, allows for the simulation of ion flux under a constant ion concentration gradient and thus a constant voltage. Simulated channel current and ion flux compare well to experimentally observed values. Animations of molecular dynamics simulations provide a better qualitative mechanistic understanding of channel formation and ion permeation and gating processes.

536089 Understanding the Molecular Physiology of the Blood-Brain Barrier Tight Junctions

Abstract:

The blood brain barrier is a selective permeability molecular barrier constituted by claudin-5 tight junction strands. Exploring treatment options for neurodegenerative diseases such as Alzheimerâ€™s remains challenging because of our poor understating of the claudin-5 tight junctions. In an effort to overcome this hurdle we use multiscale molecular dynamics simulations to study the intricate self-assembly of claudin-5 transmembrane proteins. Based on the simulation we were able to identify the stable dimer conformations that result from membrane driven claudin-5 cis interactions and their corresponding pore structures. Our results reveal the nature of the molecular interactions that drive claudin-5 aggregation in forming the tight junctions, and are in good agreement with the previous experimental predictions. The insights form this study provides novel insights into the physiology of the blood-brain barrier and will aid in designing the next generation of smart therapeutics that can permeate the blood-brain barrier.

537473 Simple Molecular Reactive Force Field - a Novel Approach to Capturing Bond-Making and-Breaking

Abstract:

A molecular-scale computational study of nucleation and growth mechanisms of materials from solution represents a challenging problem. Early-stage (small scale) nucleation can be tackled using expensive methods such as Density Functional Theory (DFT). This approach has the advantage of allowing the representation of accurate reaction mechanisms but are slow and costly to perform and hence are limited to small system sizes. Methods to accelerate such studies within a semi-empirical Molecular Dynamics (MD) framework do exist using the REAXFF and COMB force fields. However, these force fields are subject to the requirement that a parameter set must already exists for the given system of interest. These two methods are well established and work well, but they are frequently difficult to parameterize. We will discuss a new Simple Molecular Reactive Force Field (SMRFF) which captures the trade-off between atomic accuracy and ease of parameterization by redefining a reactive force field as the smooth transition between a short-range pair potential, and a long-range Lennard-Jones and Coulomb potential. This work elaborates how the implementation of a Tersoff model as a "reactive" short-range pair potential within SMRFF, with a user-friendly parameterization package, can be used to parameterize systems such as PbI2, MAI, and MAPbI3.

524780 Rational Development of Reactive Force Fields

Abstract:

We are presenting a novel workflow to optimize a reactive force-field based on existing quantum or thermodynamic data. This workflow bases on minimizing an error function against existing experimental or theoretical data in a parallel fashion using a greedy algorithm. Several implementations of reactive force-fields (REBO, ReaxFF) are discussed. We show the versatility on a variety of systems including silica gelation, carbon capture using pyrovates, and polymerizations for additive manufacturing. In all cases the system is validated against existing data.

526426 Achieving Large Scale Quantum-Accurate Reactive Molecular Dynamics: The Chebyshev Interaction Model for Efficient Simulations (ChIMES)

Abstract:

Many of the most challenging problems in chemical engineering involve reactivity in complicated systems. First principles methods like density functional theory (DFT) and density functional tight binding (DFTB) are attractive for studying these problems, however, both methods are confined to systems of a few 100s atoms, and in the case of DFT, 10s of ps. These limitations have motivated extensive work towards development of accurate, scalable, and reliable reactive force fields but these models generally rely on complicated functional forms requiring time consuming parameterization schemes. As a result, the physiochemical space over which parameters are available is limited.

In this work, we present the Chebyshev Interaction Model for Efficient Simulations (ChIMES), a new reactive force field and semi-automated fitting framework that retains most of the accuracy of DFT while decreasing computational requirements by several orders of magnitude. ChIMES models are comprised of n-body atomic interactions constructed from linear combinations of Chebyshev polynomials, and are entirely linear in fitting coefficients. Thus, model parameters can be rapidly (and iteratively) generated by force matching to short DFT trajectories. ChIMES models are particularly well suited for studies that frequently rely on costly quantum simulations for elucidation and interpretation of experiments, and to date, have been successfully applied to problems including chemical reactivity under extreme conditions and surface catalysis, which can probe time and lengths that are many orders of magnitude longer that what can be achieved with standard DFT. ChIMES is able to overcome this time and length scale gap, allowing for direct, one to one comparisons to experiment, in many cases for, the first time. Model details will be presented as will application to reactive nucleation of a carbon phase from liquid carbon monoxide.

This work was performed under the auspices of the U.S. Department of Energy by Lawrence Livermore National Laboratory under Contract DE-AC52-07NA27344. Document release number LLNL-ABS-749589

526790 Sticking Efficiency of Polyaromatic Hydrocarbons at High Temperatures By Reactive Molecular Dynamics

Abstract:

The early stages of soot formation and growth in the gas-phase involve collisions between Polyaromatic Hydrocarbon (PAH) molecules that eventually react and stick leading to a larger cluster. Simplified approaches are often adopted to model cluster collision growth rates, where the sticking rate is calculated assuming hard-sphere potential interactions between colliding entities that bind with a sticking efficiency of unity. This approach, however, neglects potential interactions between monomers and nanoclusters.

Dimerization of common PAH molecules, such as naphthalene, pyrene, coronene and pentacene, is investigated by reactive Molecular Dynamics simulations accounting for accurate potential interactions that increase the binding rate as compared to the hard-sphere approach. Furthermore, the kinetic-internal energy exchange during collision is accounted for in such simulations leading to decrease in the binding rate in comparison to conventional approaches. A framework is presented where the binding probability is systematically determined as function of relative velocity of PAH molecules, their impact parameter and temperature. The PAH dimerization rate is determined by quantifying their binding probability function and their sticking efficiency is obtained by *ab initio* trajectory calculations. Heavy PAH molecules (e.g. coronene) exhibit higher sticking efficiency compared to lighter ones (e.g. naphthalene), consistent with the literature.

The accurate determination of such sticking rates allows revisiting current coagulation rates of soot nanoparticles. The revised coagulation rates can be employed in mesoscale aerosol dynamics and soot formation models.

532093 Building Better Water Force Fields: A Systematic and Reproducible Optimization of 3- and 4-Point Water Models with an Improved Nonbonded Functional Form

Abstract:

The development of accurate molecular mechanics force fields is a significant challenge that must be addressed for the continued success of molecular simulation. We apply the recently developed ForceBalance method to automatically derive accurate force field parameters using flexible combinations of experimental and theoretical reference data. The method is demonstrated in the parametrization of two rigid water models with an alternative functional form describing van der Waals nonbonded interactions, yielding new parameter sets (TIP3P-FBV and TIP4P-FBV) that accurately describe many physical properties of water.

538829 An Accurate Force Field for Graphitic Materials Including Virtual Pi Electrons and Applications to Understand Carbon Nanotube Dispersion in Solvents and Polymer Solutions

Abstract:

A new, compatible all-atom force field for graphene, graphite, carbon nanotubes and related materials is introduced as part of the Interface force field (IFF) that enables improvements in the reproduction of surface energies, hydration energies, contact angles, cation-pi interactions, and prediction of multiphase assembly by an order of magnitude compared to earlier models. The models are compatible with IFF, PCFF, CVFF, CHARMM, and several other force fields. Molecular dynamics simulations with these interatomic potentials that include virtual Ï€ electrons are reported to explain the interaction of pristine single wall CNTs with solvents and polymer solutions at 25 Â°C. Debundling and dispersion of carbon nanotubes (CNTs) in polymer solutions plays a major role in the preparation of carbon nanofibers due to early effects on interfacial ordering and mechanical properties. A roadblock toward ultrastrong fibers is the difficulty to achieve homogeneous dispersions of CNTs in polyacrylonitrile (PAN) and poly(methyl methacrylate) (PMMA) precursor solutions in solvents such as dimethyl sulfoxide (DMSO), N,N-dimethylacetamide (DMAc), and N,N-dimethylformamide (DMF). The results quantify CNT-solvent, polymer-solvent, as well as CNT-polymer interactions in atomic detail, including wrapping of PMMA around CNTs which remarkably changes interfacial properties. Quantitative comparisons with solubility parameters and experimental data are reported. See details in: Pramanik et al. ACS Nano 2017, 11, 12805-12816.

538441 A Hybrid Approach Toward Systematically-Derived Implicit-Solvent Coarse-Grained Lipid Models

Abstract:

Biological membranes are intimately involved in many cellular processes, including cell signaling, energy storage, compartmentalization, and active transport. These membranes are composed of a diverse array of phospholipids, which self-assemble into bilayers due to their amphiphilic nature. Many of the macroscopic properties of these membranes, and resultant insights into their biological functionality, would benefit from a molecular understanding of their behaviors. To this end, implicit-solvent coarse-grained (CG) models enable investigation into molecular dynamics at biologically-relevant length- and time-scales. However, as water is largely responsible for the entropy-driven hydrophobic effect that underlies lipid behavior, parameterization of effective CG interactions has been notoriously difficult. In this talk, we will present an approach to systematically derive high-fidelity, low-resolution CG lipid models from all-atom simulations by leveraging two techniques: multiscale coarse-graining (MSCG) and relative entropy minimization (REM). We first provide a comprehensive comparison between MSCG- and REM-derived CG lipid models, highlighting their respective benefits and deficiencies. Then, in contrast to conventional methodologies, we demonstrate the importance of featurizing interfacial water in the form of virtual CG particles; here, effective interactions are derived from a hybrid MSCG/REM framework. The resultant CG models recapitulate critical physical phenomena, including robust self-assembly with diverse morphologies and phases, which we attribute to a redistribution of entropic and enthalpic driving forces. As a final note, we discuss the potential to generalize our approach for biomolecular CG models with improved transferability.

523200 Improved Directional Hydrogen Bonding Interactions for the COSMO-SAC Model for Prediction of Activity Coefficients

Abstract:

A Novel Approach for the Determination of Directional Hydrogen Bonding Interactions for the COSMO-SAC Model based on Molecular Electrostatic Potentials

Chun-Kai Chang [a], Wei-Lin Chen [b], David T. Wu [c] and Shiang-Tai Lin [a*]

[a] Department of Chemical Engineering, National Taiwan University, Taiwan

[b] Daxin Materials Corporation, Taiwan

[c] Department of Chemistry and Department of Chemical and Biological Engineering, Colorado School of Mines, Golden, CO, USA

ABSTRACT

The COSMO-SAC model is a powerful tool for predicting the phase behavior of liquids based on quantum chemical calculations of molecules solvated in a medium of infinite dielectric constant. However, many studies have shown that its accuracy is reduced for mixtures containing hydrogen bonding species. Therefore, there has been a continuous effort to improve the description of hydrogen bonding interactions. In a recent work, Chen and Lin showed that the consideration of directional hydrogen bonding in the COSMO-SAC model significantly improves the description of properties of associating fluids (Phys. Chem. Chem. Phys., 2017, 19, 20367-20376). In their method, the direction of a hydrogen bond was determined based on the VSEPR theory; however, this geometric approach does not reflect the local electronic environment of the lone pairs and cannot be unambiguously applied to certain molecules such as CO_2 and HF. In this work, we adopt a new scheme that determines the hydrogen bond acceptors of a molecule based on the minima of the molecular electrostatic potential (MESP) instead of the molecular geometry and bonding assignment. This method is applicable for all types of compounds as the needed electrostatic potential can be obtained from quantum chemical calculations. In this work, we compare the performance of the newly proposed method, referred to as COSMO-SAC(DHB)/MESP, to the VSEPR based method, denoted as COSMO-SAC(DHB)/VSEPR, and the original COSMO-SAC(2002) and revised COSMO-SAC(2010) models. The data considered include vapor-liquid equilibrium (VLE) of 586 binary pairs, liquid-liquid equilibrium (LLE) of 190 binary pairs, 431 infinite dilution activity coefficients (IDAC) and 291 octanol-water partition coefficients (K_{ow}). It is shown that the directional hydrogen bond interactions determined based on the MESP provides the best overall accuracy for the systems studied.

Keywords: COSMO-SAC, Directional Hydrogen Bond, Molecular Electrostatic Potential.

AUTHOR INDEX

AUTHOR INDEX

AUTHOR INDEX

AUTHOR INDEX

AUTHOR INDEX

AUTHOR INDEX

AUTHOR INDEX

AIChE
120 Wall Street, FL 23
New York, NY 10005-4020

ISBN 978-1-5108-7615-6

9 781510 876156

CURRAN ASSOCIATES INC.
proceedings
.com